Trust No One

Trust No One

Inside the World of Deepfakes

Michael Grothaus

HODDER*studio*

First published in Great Britain in 2021 by Hodder Studio
An Hachette UK company

1

A CIP catalogue record for this title is available from the British Library

Hardback ISBN 9781529347975
Trade Paperback ISBN 9781529360288
eBook ISBN 9781529360301

Typeset in Bembo by Manipal Technologies Limited

Printed and bound in Great Britain by Clays Ltd, Elcograf S.p.A.

Hodder & Stoughton policy is to use papers that are natural, renewable
and recyclable products and made from wood grown in sustainable forests.
The logging and manufacturing processes are expected to conform
to the environmental regulations of the country of origin.

Hodder & Stoughton Ltd
Carmelite House
50 Victoria Embankment
London EC4Y 0DZ

www.hodder-studio.com

For my dad.

CONTENTS

Prologue

Where's All the Chaos?

On Tuesday, 3 November 2020, shortly after 11 p.m. on the East Coast of the United States, the first network declared the winner of the 2020 presidential election. President Donald J. Trump had been reelected for a second term.

The call from the first network was followed by a similar call from another network minutes later. By quarter to midnight, all three major national broadcast networks and the three major cable news networks declared President Trump the winner, beating Democratic hopeful Joe Biden by the narrowest of margins in critical swing states, including Arizona, Florida, Minnesota and Pennsylvania.

The outcome of the 2020 election was a devastating blow to those on the left hoping that the immense suffering 2020 had brought, with the global Covid-19 pandemic leading to historic deaths and unemployment in the country, might be enough to convince the all-important undecided voter that a change in leadership was needed. And throughout the summer and into the fall of 2020, that hope seemed justified. According to most polls, Biden led by an adequate margin in critical swing states. Despite the fact that the country had never been so polarised, it looked like his appeal to undecideds was enough that, provided no significant missteps,

Biden simply needed to coast to the 3 November finish line to win it.

But everything changed with that late October surprise.

The video first appeared on Facebook ten days before the election. It was posted to the page of a small group of Pennsylvanian deer hunting enthusiasts, before being copied and shared among other, larger, hunting groups on the platform. From there, the video made its way onto the page of a popular Second Amendment Facebook group, where it exploded. Within the hour, it was shared over half a million times. By the end of the day, after being reposted to Twitter, the video had racked up close to twenty million views.

The footage was damning – all twelve seconds of it: it showed Joe Biden, clearly in frame, surrounded by a huddle of advisors. It was obvious one of the group was recording the meeting without the knowledge of the presidential-hopeful. In the Dutch-angled clip, a tough, steely-eyed Biden glared at someone out of frame. Giving a grim nod, Biden said the words that would ultimately cost him the election: 'America's gun violence is too great. The day I'm sworn in, I'm signing an executive order confiscating and banning all firearms.'

The video was what media and cybersecurity experts had long feared. And it was, of course, fake. A deepfake, to be precise. Created, virtually out of thin air, by artificial intelligence. Yet, despite its insincerity, it was good enough for its intended purpose. It was almost 24 hours before the first media outlet confirmed its inauthenticity. Additional media outlets followed the same day. Yet still, other media, those more sympathetic to President Trump's reelection campaign, having already reported on the faux scandal, never followed up with clarification that the video being shared on social media, and in texts and private WhatsApp groups across the country, was indeed fake. And, by the time of the election the following week, that adequate margin Biden had led with throughout autumn evaporated by election night as enough undecideds, still believing the deepfake real, flipped to Trump, and pro-gun advocates turned out in record numbers.

Now, none of what I've described so far, thankfully, *actually* happened. There was no deepfake of Joe Biden that went viral and cost him the election. The Democratic contender did have a hard-fought campaign, but he came out victorious. And despite Trump, his enablers and a group of insurrectionists trying to stop Biden from being sworn in by almost all means necessary, Joseph R. Biden did become the 46th president of the United States at noon on 20 January 2021.

Yet, in the years running up to the 2020 election, you could be forgiven for fearing that a deepfake could very well alter the course of history. There certainly was no shortage of headlines warning of such an impending calamity. Just take this selection from various publications:

'Deepfakes threaten the 2020 election.'[1]
'Will Deep-Fake Technology Destroy Democracy?'[2]
'Putin developing fake videos to foment 2020 election chaos: "It's going to destroy lives."'[3]
'Fake news is about to get so much more dangerous.'[4]
'The 2020 campaigns aren't ready for deepfakes.'[5]
"Nightmarish': Lawmakers brace for swarm of 2020 deepfakes.'[6]
'Top AI researchers race to detect "deepfake" videos: "We are outgunned."'[7]

'Deepfakes are coming for American democracy. Here's how we can prepare.'[8]

'"Deepfakes" called new election threat, with no easy fix.'[9]

'Fake videos could be the next big problem in the 2020 elections.'[10]

'US Intel agencies warn about Deepfake video scourge.'[11]

'The sinister timing of deepfakes and the 2020 election.'[12]

'Deepfake Videos Set to Wreak Havoc.'[13]

Pretty alarming, right? But the thing is none of what those headlines warned of – all written between 2018 and 2020 – came to pass. Deepfakes didn't threaten the 2020 election. They didn't destroy democracy. And Putin did not, as far as we know, release a single deepfake that caused chaos. In short, there was no 'nightmarish' deepfake 'havoc' as these headlines portended and cybersecurity experts feared.

So, what happened? Or rather, what didn't happen?

It's an important question with a complex set of answers, which is no surprise because everything about deepfakes is complex (except, unfortunately, for the expertise needed to create them).

Those headlines, while alarmist, certainly didn't warn of scenarios that were outside the realm of possibility. As a matter of fact, the only thing we can say with certainty about any of those articles is that what they warned of hasn't come to pass – yet.

But things are definitely heading in that direction.

In fact, we are in the middle of emptying out a great big Pandora's Box of this novel technology, with its reality-altering powers. I say 'Pandora's Box' because while deepfake technology is one that can – and already has been used to – cause great personal harm, it is also a technology that will extraordinarily impact everything from politics to entertainment to healthcare – and perhaps to even death itself.

See? I told you it was complex.

And that complexity – that extraordinary, terrifying and, frankly, awe-inspiring complexity of deepfakes, those creating them, and those whom the technology has already impacted and will impact (including myself) – is what this book explores. It's an exploration of the very real, *human* side of this fabricated wonderland. And it's an exploration that will take us from Phoenix, Arizona to Mumbai, India to Ibaraki Prefecture in Japan, to some of the darkest diasporas on the internet. In the process we'll meet everyone from a queer Muslim filmmaker to a beautiful motorcyclist with a head of hair beloved by all, to a secretive celebrity pornographer for hire and a successful eco-conscious fashion influencer with some godlike abilities.

The common thread between them?

Deepfakes, of course.

Chapter One

The Forger and the Inspector

At its most basic, a deepfake is media – usually video, but it can be only audio, too, or a combination of both – that has been altered using artificial intelligence and machine learning techniques. This altered media deviates from the original to depict an event that, in reality, never happened – and its only limitation is the deepfaker's imagination. You could make a deepfake that shows something clearly fantastical, such as Martin Luther King, Jr dunking in place of Michael Jordan in the final game of the 1997 NBA Championships – 29 years after the civil rights leader was slain. It could be for a bit of light entertainment, like making Lady Gaga wish your sister a happy birthday on WhatsApp. Or, much more worryingly, you could create something that is meant to be passed off as a factual event. An event showing, say, a presidential hopeful saying something that will irreparably damage their election campaign. It's easy enough to imagine the delightful series of videos you could conjure up at your whim – and I do mean you: this technology is easy enough to get hold of and even easier to use. But it is in that third scenario – the deepfakes specifically designed with ill intent to trick people into thinking the video playing before their eyes is authentic – where the danger lies. But we'll get into that later.

Before the advent of the technology that underlies deep-fakes, any of these examples could have been created using the traditional CGI techniques we're all so familiar with that Hollywood has used in its tentpole blockbusters for almost three decades now. Yet, such conventional techniques to manipulate video – even ten seconds of altered footage – have always relied on massive amounts of money, time, and, most importantly, talent in the form of a team of skilled artists proficient in animation and computer graphics. Creating something that was believable, even unidentifiable from reality was, for the most part, out of reach to you or I.

Enter deepfakes.

The power of deepfake technology lies in its ability to pull off ultra-realistic alterations to authentic media in a fraction of the time standard CGI takes, at virtually no cost, and without any artistic or computer graphic skills required. In fact, all you need to create a deepfake is the right software – which is freely available on the internet – and access to a computer with a moderately powerful graphics card, which you can rent remotely via the cloud right from your cheap $300 laptop – no dedicated purchase needed. Throw in a video that you want to alter and have enough images of the person you want to deepfake into the video, and the software will blend everything together for you. And then *voila*! You've just created a realistic deepfake. Though it sounds so simple, perhaps it is this ease with which deepfakes can be created that perfectly exemplifies what makes them so impressive – along with their massive potential to be all at once terrifying, fascinating, and awe-inspiring. Which is what, of course, drew me in to it in the first place.

Deepfakes seem, on the surface, like magic – or black magic, perhaps. But a clue to the underlying technology that enables

deepfakes lies right in the 'deepfakes' name itself. It's a port-
manteau – derived from the term 'deep learning' and the word
'fakes'. While the 'fakes' part is obvious, the 'deep learning'
part is what hints at what is truly going on beneath the surface:
artificial intelligence.

Deep learning is a type of machine learning – one of the
main techniques that powers artificial intelligence. Deep learning is a vital tool in the field of computer vision, where computer systems are tasked with identifying subjects in videos and
photos and is used in everything from self-driving car and facial
recognition systems to the apps on your smartphone that automatically tag your friends in pictures.

While traditional machine learning uses layers of algorithms
to process data in order to automatically carry out specific tasks
using that data (such as identifying people or objects in videos)
and can become more proficient at carrying out those tasks
over time with guidance or tweaks to its algorithms from a human engineer, deep learning is a more advanced form of machine learning that gets better at a task by teaching *itself* how to
improve – how to tweak its own algorithms – without involving a human in the process. Thanks to this deep learning ability,
a self-driving car's computer systems or a facial recognition
system gets better on its own at identifying people and objects
with the more of them it sees. Given how remarkably powerful
this technology is it's no wonder that storylines of rogue AI
systems litter popular entertainment. After all, the idea that you
can leave a machine to its own devices – just set it on its way
and trust it to do what it concludes is best – is, well, unsettling.

Without going too in-depth into the tech side of things, deep
learning accomplishes its self-improvement on the deepfakes
front, essentially, by using tools known as generative adversarial

networks (GANs), designed circa 2014 by a 29-year-old computer scientist named Ian J. Goodfellow and his team of researchers at the Université de Montréal.[14] A GAN is a framework that pits two artificial neural networks (ANNs) against each other. For our purposes of understanding deepfakes, these two ANNs can each be considered an individual AI. The GAN, then, pits one AI in a game against the other. In the runup to that game, each AI is first given access to the same dataset. In the case of deepfakes, that dataset is a collection of authentic photos of a person about whom a deepfaker wants to create a falsified video. Both AIs review this dataset of authentic photos – the more photos, the better – which teaches, or 'trains', each AI what a real photo of that person looks like.

Then the game begins: one AI is given the role of a forger, and the other the role of an inspector. It's then the job of the forger AI to create a fake photo from scratch – a deepfake – of the subject, based on what it knows of the person from the dataset of the authentic photographs, that can fool the inspector AI into thinking it's a real photograph of that person.

Once the forger AI has created its first forgery, it shows this photo to the inspector AI, who then needs to mark it as either authentic or fake. If it's the first forged photo – or even among the first thousand forged photos – the inspector AI should easily be able to tell the photo the forger AI is showing it is a fake. But here's where the magic of the self-learning among deep learning systems comes in: every time the inspector AI marks a forged photo as fake, the forger AI reviews the aspects of that forgery, which naturally wasn't good enough to fool the inspector AI, looks back over the dataset of the authentic training photos, identifies what's different between the two, and then tries to fool the inspector AI again with another, improved forgery.

This process of attempted trickery is played out thousands and thousands of times. And with each forged photo rejected, the forger AI only gets better at knowing where it needs to improve. As it improves, the fake photos it generates become more realistic until, ultimately, the forger AI finally generates a fake photo that tricks the inspector AI into thinking it's the real deal – that is, thinks it's a genuine photo from the original training dataset. It is when this happens that the forger AI can realistically create a photo – a deepfake – of the subject in the original dataset. The game has been won. And from there, the forger AI has the skills to create as many realistic images of the person as needed and can then easily apply their likeness onto a video (which is essentially a series of still photographs), showing that person doing whatever the deepfaker wants them to.

It's important to note that, in its genesis, a deepfake requires both a trained AI that can generate thousands of forged photos of the subject's face in an almost infinite range of possible expressions, as well as an original video of someone else that that forged face can then be overlaid and mapped onto. Current technical limitations do not allow deepfake software to generate *everything* seen in a deepfake video, such as the person's entire body or their surrounding environment, which is why an original video is required to act as a canvas for the deepfaker. For now, at least.

In our Martin Luther King, Jr scenario, for example, deepfake technology, and general computing processing power, doesn't yet have the capabilities to easily generate everything in our scene from scratch: the basketball court, the screaming fans, the other players, the ball and an MLK with Michael Jordan-like abilities. Instead, a deepfaker would first select a real video of the actual Michael Jordan dunking, then train deepfake software on a dataset of actual images of Martin Luther King, Jr and the Michael Jordan from the video. The deepfake software would then overlay the forged imagery of MLK's face onto Michael Jordan's face in the original video – including matching the fake MLK's expressions frame-by-frame to that of the real Michael Jordan's expressions to give our synthetic MLK physical authenticity based on the context of the scene.

But it's also important to note three other things about the capabilities of deepfake software. First, within the decade, computing power and artificial intelligence will have advanced so much, deepfake software will most likely be able to easily render completely fabricated, photo-realistic scenes of anyone doing anything anywhere without any real-world footage needed as the canvas to manipulate. I find this as exciting as it is frightening.

Second, such capabilities aren't needed for deepfakes to have a staggering impact on our world right now. The current ability of deepfake software to easily create realistic faces and map them onto subjects in existing video is more than enough to create falsified media that can pass as authentic today – and, potentially, wreak havoc in people's lives, as we'll soon see.

And third, though current deepfake software is more than capable of creating realistic faces that can be mapped onto another person's body, many such deepfakes don't even need to alter an entire face. A deepfake video where a deepfaker wants a politician to be caught saying something they, in fact, never said, as our earlier example shows, only requires authentic existing footage of the politician. Because that politician already has the face the deepfaker wants, the deepfake software can simply be tailored to change the politician's lips and underlying facial movements. Combine those subtle changes with deepfaked audio of the politician's voice and you have the fabricated video you wanted without having to recreate the target's entire face anew. Imagine all the things a video like that could be used for.

At this point, you may be thinking you're not the type who would ever use deepfake software – that all this stuff I'm talking about is so far away from you and your life. But the reality is that you (or your children, at least) may have already been using the technology for years. Almost as soon as Goodfellow and his team published their GAN work, others in the tech sector began iterating on or creating similar computer vision processes to be used in their own products. One of the first consumer applications of deepfake technology was seen in an app released just the following year, in December 2015.

Face Swap Live[15] was the first mainstream smartphone app that launched the craze and allowed smartphone users to swap faces with friends or family members in real-time by pointing their smartphone's camera at themselves. The app allowed them to record these face swaps and then post short clips of the spectacle to social media. Despite the relatively low quality of the face swaps compared to today's deepfakes, such clips soon went viral, and were even showcased on morning talk shows. This was effectively the first time in history the general public became acquainted with deepfake technology – and many couldn't get enough of the oddly funny videos. This included an old college friend of mine who had worked as a VFX specialist in Hollywood for close to fifteen years at the time. Over the holidays, just after Face Swap Live launched, he texted me a face swap involving him and his wife.

'Have you heard about this new app?' he wrote alongside the video featuring their smiling transplanted faces. 'It's AI, apparently.'

I told him I had heard about it, and it was a great example of the maturing field of computer vision. But I don't think my friend thought there was anything 'mature' about the technology – not based on the slightly gelatinous look of the swapped faces in the video he'd sent me.

'I think my job is safe for now,' he replied along with a laughing emoji.

Well, the 'for now' part was right.

Shortly after the deepfakes created by Face Swap Live users went viral, other, more prominent companies moved into the burgeoning face swap arena. No less than two months after Face Swap Live debuted on the Apple and Google app stores, social media giant Snap introduced the 'FaceSwap' lens into its Snapchat app, in February 2016. Their lens worked almost identically to Face Swap Live but gave Snapchat users the ability to create and post face swaps from within the Snapchat platform itself without requiring a third-party app. Given Snapchat's base of, at the time, over 150 million young, plugged-in, tech-savvy users, Snapchat's inclusion of its FaceSwap lens propelled deepfake technology to new heights.

From the United States to Nigeria to Japan, and back again, face swaps swept across the globe like never before thanks to these two apps. It was almost impossible to view my social media feeds without seeing the fabricated videos. Everyone from tech blogs to respected publications like *The Washington Post*[16] were writing about the new craze – I was writing about the new craze. Needless to say, the world's biggest celebrities and influencers soon joined in on the trend, too, posting their own face swaps online and garnering millions of likes in the process. And, by the end of 2016, barely a year after Face Swap Live debuted,

Apple announced it was the second best-selling paid app on the App Store.[17]

The world had fallen in love with deepfakes without ever even questioning the technology behind them.

But while the face swap craze was all the rage on smartphones in 2016, by 2017, an entirely new breed of deepfake creator popped up, bringing with them even more evolved types of deepfake content. Far from just your average smartphone user swapping their own faces, these creators were individuals who took the time to learn about the underlying GAN technology and taught themselves to use burgeoning desktop deepfaking software to train their own datasets, allowing them to swap the faces of not themselves, but Hollywood actors, into famous movies they'd never starred in.

Soon YouTube was packed with dedicated channels maintained by content creators uploading the 'fancasting' deepfakes they created. There've been fancasts of Nicolas Cage deepfaked into the role of Neo in *The Matrix* (sorry, Keanu), Michael Jackson replacing Jonny Depp in *Charlie and the Chocolate Factory* and Tom Cruise replacing Robert Downey, Jr in *Iron Man*. Other deepfake fancasts included Jim Carrey replacing Christopher Lloyd as Doc Brown in *Back to the Future* and a young Harrison Ford replacing his real-life replacement, Alden Ehrenreich, in the *Star Wars* prequel *Solo: A Star Wars Story*.

Speaking of *Star Wars* – and underscoring just how rapidly deepfaking software had advanced in the few short years since the release of Face Swap Live – in January 2018, a popular YouTube deepfaker, by the name of Derpfakes, posted a deepfaked clip he made of Princess Leia from the 2016 film *Rogue One: A Star Wars Story*. While a box office hit, one of the main complaints about *Rogue One* was that although Disney spent months and millions on traditional CGI techniques to place the digitally replicated face of a young Carrie Fisher onto a stand-in for one of the final scenes in the film, the end result came off looking . . . not quite right. The filmmakers hired Norwegian actress Ingvild Deila[18] whose facial features – her chin and cheekbones – were almost identical to that of Fisher's when she shot *Star Wars* in 1976. They painted Deila's face with motion-capture dots and painstakingly shot the new scene from precise angles. Lucasfilm's Industrial Light & Magic (ILM) then took over. The expert team of computer artists, including modellers, animators and digital texture and lighting experts, worked night and day to create a computer-generated digital replica of Fisher's face to overlay onto Deila's.

Though the cutting-edge CGI techniques ILM used ostensibly ticked all the correct boxes, it took a lot of time, money and talented manual labour to get a young Fisher back into the role of Princess Leia. Yet despite all that, the digitally recreated Leia's final on-screen presence still made many in the audience, myself included, feel disturbed. There was this sensation that we couldn't shake, commonly known as the uncanny valley effect, the disconnect between what we see and what we feel when we look upon something that is designed to look natural but isn't. Somehow, our minds know that something about the superficially natural-looking thing is not quite right. As for the CGI'd Fisher in particular, there was something about the face – an eerie lifelessness embedded within its realistic facade – that, far from being enthralling, was simply unnerving.

That's where Derpfakes's YouTube clip comes in. Created just thirteen months after the release of *Rogue One* – and using freely available desktop deepfake software – Derpfakes deepfaked the real, young Carrie Fisher's face onto ILM's computer-generated face. The results were unexpected – and impressive. Derpfakes's deepfake of Fisher appeared more natural-looking and didn't produce the same uncanny valley effect – astonishing considering all the tech, time, and money Disney had thrown at professional CGI artists to bring a young Fisher back to the silver screen. As Derpfakes noted in his deepfaked clip's description, 'keep in mind this fake was done on a standard desktop PC and completed in the time it takes to watch an episode of [*T*]*he Simpsons*'.[19]

How easy would it have been for you, or me, to do the same?

It turns out, incredibly so. Unlike the software developer or hacker tropes that we're all familiar with, a deepfaker could literally be you, me or your average person working in the café opposite you. While the code underlying deepfake software might not be decipherable without an advanced understanding of computer science, deepfake technology's power lies in the fact that anyone can wield it regardless of their computing comprehension. It's this universal accessibility that has always been a fundamental promise of artificially intelligent tools. And while some might find that unnerving, there's nothing inherently scary about the idea that technology is available and accessible to most. In fact, I'd say there's something almost utopian about that idea. After all, no technology is inherently bad. That's just a fact.

Yet when you consider the scope for what could be achieved and look back over history at the various instances – and there are sadly far too many to name – where technology has been perverted and used for cruel ends, suddenly, the idea that anyone can wield deepfake technology sounds less ideal and more sinister. And it cannot be denied that deepfake technology can certainly be used to do bad stuff, as we'll see very soon, even when there isn't any conscious intent to do harm.

As desktop-powered Hollywood face swaps were becoming all the rage on YouTube, deepfake face swap technology on smartphones wasn't letting up either. As face-swapping apps grew in popularity, their numbers – and the quality of their deepfake powers – also rapidly improved.

In January 2017, a new deepfake-powered app hit the Apple App Store, followed by the Google Play Store a month later. Called FaceApp, the underlying technology powering the app was essentially the same as that powering Face Swap Live: it used artificial intelligence-powered neural networks to alter a user's face. However, instead of swapping the user's face with another's, FaceApp's tech allowed users to alter certain characteristics of their own face. After uploading a selfie via the app, users could then select from tens of filters capable of changing their appearance. Such deep learning-powered filters enabled FaceApp users to age their selfie up in years to see what they might look like as an elderly person, de-age themselves to look younger, alter their hairstyles, and even change their gender. Keep in mind these filters were not simply layering special effects on top of the uploaded selfies, they were using complex deep learning processes to actually alter the selfies pixel-by-pixel. Other filters allowed users to tweak just one aspect of their uploaded photo, for example, turning a frown or a neutral set of lips into a big smile.

This was deepfaking for self-improvement.

But with the ever-increasing popularity of deepfake-powered apps came the first signs of controversy.

In April 2017, just months after FaceApp launched, users noticed something about one of the app's filters. Originally dubbed the 'hot' filter, when applied to a selfie the user uploaded, it would use its complex deep learning algorithms to make the person look, ostensibly, more attractive. However, what users soon noticed was that this 'hot' filter, when applied to selfies of people of colour, lightened their skin to make them look whiter. Additionally, it gave them more European-like features by making other deep learning-powered alterations to their face, such as narrowing the user's nose. Needless to say, the app was soon accused of racism.

FaceApp's developers moved quickly to address these claims, saying the whitening and European-ising of dark-skinned individuals were not intentional; that it was, in fact, the fault of the deep learning underpinning FaceApp's software. Now, usually a developer blaming racist behaviour on the app's code and not themself would be laughable. It would be like a writer blaming the pen for a racist rant. But in this case, it happened to be somewhat true.

Unlike traditionally coded apps, where developers write every line of code that specifically instructs an app what to do, apps that rely on deep learning, as we've seen, teach *themselves* what to do. FaceApp's creators didn't instruct the app to lighten dark skins – the app taught itself to do so because of the study materials its neural networks used. In FaceApp's case, at the time, its software had been taught with images of predominantly light-skinned people; the developers had neglected to use a diverse training set from which the AI could learn – and this, of course, had the unintended effect that the app 'believed' that the norm among all people was lighter skin – because that's the shade of skin most people had in its training set. Because of that, FaceApp's deepfake software thought that to make someone hotter – to make them ideal – you needed to make their skin lighter, too, since, to FaceApp's deep learning algorithms, that's how skin was *supposed* to look.

This 'training set bias', as FaceApp's founder and CEO called it at the time, isn't limited to FaceApp or even deepfake software in general.[20] Any computer vision systems that rely on deep learning can be inadvertently fed racial biases based on the image datasets used to train them. There have been multiple cases of facial recognition software used by police, for example, that has been found to be biased against Black people, returning more 'false-positive' results for persons whose skin is darker.[21] This is likely because the datasets used to train that facial recognition software does not have enough racial diversity. Though the intention of racism wasn't actively coded into the software, the software works as trained. The people who provide the datasets that train the software need training themselves in selecting more inclusive and representative datasets to begin with.

After the 'hot' filter uproar, FaceApp quickly altered the filter to exclude the lightening of the skin and renamed the filter 'spark' so as to not imply any positive connotations to the deepfaked changes made to the selfies. Shortly after that, the spark filter was removed from the app entirely. Yet this incident hasn't stopped various iterations of essentially the same problem popping up time and again across a variety of apps and filters. Human judgement has always been, and will always be, occasionally suspect. Particularly when things are moving at such a fast and, dare I say, exciting pace, which causes us to act first and consider the consequences later.

I've written about the tech industry and its effects on our world for long enough to know that there are some years where technological advancements seem to move along at a glacial pace. Yet there are other times where a technological breakthrough is so revolutionary it seems to catapult us ten years into the future in a matter of hours (the commercial internet and the release of the original iPhone are two such instances that come to mind). For individual apps, this rapid advancement is even rarer – but it happens. Yet such advancements – no matter how well-intentioned – can lead to unexpected suffering.

In February 2020, a new deepfake app called Reface[22] hit the app stores. Like its predecessors, Reface relied on deep learning technologies to deepfake users' faces onto other people. However, Reface was leaps and bounds ahead of traditional smartphone deepfake-powered face swap and filter apps. This was for two crucial reasons.

First, Reface took a cue from the popularity of the fancasts deepfake content creators posted to YouTube of Hollywood actors starring in films they never did. Seeing how frequently these altered movie clips went viral, the Reface team likely realised their app could be a hit if it could allow users to deepfake their faces over the faces of actors in popular movies – fancasts starring you as Iron Man instead of Robert Downey Jr (or fancast-favourite Nicolas Cage).

And this is precisely what Reface does. Users can insert their own face over the faces of actors in clips from popular television shows like *Game of Thrones* or Marvel's films and other Hollywood blockbusters, as well as over the performers' faces in World Wrestling Entertainment matches, and even over the faces of musicians in their hit music videos. Thanks to its deepfaking tech, Reface lets its users be A-listers, pro athletes, and Billboard music stars – no acting, athletic, or musical talent (or computer graphics talent, for that matter) required. No wonder the app's tagline is 'be anyone'.

The second reason Reface was leaps and bounds beyond other smartphone deepfake apps has to do with the first: the video clips. Before Reface, in order to deepfake a person into the role of a character in a movie, as YouTube fancasting deep-fakers had been doing for years, you needed desktop deepfake software and a moderately powerful graphics card, along with a large dataset of the face of the person you wanted to deepfake into the video (such as hundreds of images of Johnny Depp's face if you wanted to insert him into Hugh Jackman's Wolverine role in Marvel's *X-Men*). The amazing thing about Reface was that the app didn't require large datasets of the person to be deepfaked into the movie clip – it just required a single photo of the user – and *bam!* – they're suiting up as Wolverine.

This staggering capability is both a testament to Reface's team of machine learning engineers as well as to how far deep learning techniques had advanced in a few short years in general. In 2016 you needed at least hundreds, if not thousands, of photos of a person to train deepfake software on so it could reach a level of forgery good enough to insert them into a falsified video realistically. By 2020, you just needed a single selfie taken with your smartphone. Think of where we'll be by the next election cycle . . .

And the thing is, Reface was accomplishing these single-photo-dataset deepfakes refined to a level of quality that would have been simply unbelievable just four years earlier. Though revolutionary at the time, Face Swap Lives' deepfakes from 2016 often looked blurred or pixelated – sometimes a face swap even looked like half-baked cookie dough with melted eyes and a mouth pasted on. No wonder my Hollywood VFX specialist friend dismissed them. Yet by 2020, Reface boasted the creation of 'hyper-realistic' deepfakes . . . with just one selfie.

And because of this stunning evolution, once again, the world went crazy for deepfakes. Reface's ability to easily and realistically swap your face into not just a random video but a *Hollywood blockbuster* brought a new wave of attention to deepfake tech – along with tens of millions of app downloads by users from all walks of life – including many of the very celebrities the app let users deepfake their own faces onto. No wonder the company proudly boasted, 'Britney is Refacing, Snoop is Refacing, Miley, Justin. It looks like the whole world is obsessed with the thing we created.'[23]

Indeed.

But with this new deepfake hyper-realism Reface delivered came something more sinister – more so than the digital racism controversies of previous apps. And this sinisterness makes those headlines we saw earlier seem less hysterically alarmist and more prophetic. To understand what I mean by that, we need to explore what happened when Reface's deepfake capabilities were used to intentionally harm another human being. In other words, it was *weaponised*.

The first time I met an openly gay person was in 1998, during my junior year of college. I was 21 at the time, and the fact I hadn't met an openly gay person until then can be attributed to both the era I grew up in and where in the country I grew up. I was born and raised in the Midwest, which was then – and still is – one of the United States' more conservative and religious regions. Though the gay rights movement in the United States had been going on since the 1960s, predominantly on the coasts, an openly gay person in the 1990s' Midwest was still a rarity – at least in the suburban circles I had been slotted into.

We became good acquaintances, friends even, always sharing a beer and a conversation at whatever dormitory party we happened to bump into each other. We were admittedly an odd duo – different sexualities, different majors, different economic and political backgrounds. Not the type of people you'd think would have much genuine interest in what the other had to say – especially at that age. But our bond began when we discovered we had a sort of shared history: that of bullying.

My friend had come out years earlier during his senior year of high school (something exceptionally courageous at the time and for his age – it was virtually unheard of back then). But almost immediately, the bullying started. In the halls of his high school, he went from being celebrated – having been a star jock on the football team – to enduring subtle, and at many times, not-so-subtle calls of 'fag', 'homo' and 'dick muncher' while going from one class to the next. Virtually no one sat near him in the cafeteria any more; people laughed at him – behind his back, when not openly to his face. Conversely, the cheers from his teammates when he scored a touchdown were quieter from then on. Truthfully, he confided to me years later, he regretted coming out that early because of what it had cost him.

Me, on the other hand, I had nothing to come out about. I spent my first three years of high school being known as the fattest kid in school – not something you could exactly hide, though I would have loved to have been able. So, it was my size that brought the behaviour my friend described. I was called names, picked on, ostracised. I ignored it the best I could, waiting in anticipation for the final bell of each day to ring so I could escape my torment till the next morning. I'd never spoken to anyone about the bullying before, until I met this friend years later in college, so part of me had believed that my suffering was as unique as my size had been. But finally speaking to my friend about our experiences, I realised that, though the reasons someone is bullied are often vastly different, the type of suffering one feels – the humiliation, the shame – is universal.

But the funny thing is, as I've gotten older and technology has advanced, I've often found myself oddly grateful for the time period in which I was bullied. In the 1990s, the bullying ended with the school day. Today, with the internet and smartphones, and social media, bullied people often have no respite from the abuse. Bullying has gone digital in the twenty-first century, and it can be relentless, following you everywhere, attacking you at any time.

Now, thanks to deepfakes, it can be more egregious than ever.

Faraz Ansari is a 34-year-old writer and filmmaker from Mumbai, India. Like my friend from college, Ansari was born gay, grew up gay and came out as gay when they were young – and they were relentlessly bullied for it (Ansari identifies as non-binary and prefers gender-neutral pronouns).

Arguably though, Ansari probably had it even worse than my college friend did. By the mid-90s, when my friend came out, many social norms in the United States had been moving in the right direction for years. While it was still unheard of for most teens to voluntarily come out back then, especially where I grew up, the United States' acceptance of LGBTQ+ peoples was advancing, however slowly.

The same could not be said for the culture Ansari grew up in. Not only was Ansari raised in India, which until as late as 2018 still criminalised homosexual acts (as of the time of writing, the country still does not recognise same-sex marriage), but they were also raised among the Islamic faith, which, historically, has seen homosexuality as a grave sin. In some Islamic-majority countries (which India is not), homosexuality is punishable by death.

'I went to an all-boys school in Bombay [Mumbai],' Ansari recalls, 'and, you know, there's this constant sort of laughter that we as queer people keep hearing in the background every time you pass by, every time you walk inside the room there's this background laughter that one just keeps hearing over and over and over again.'

Tales of those old taunts Ansari received growing up as an openly gay Muslim in India – it's sadly the type of story I've heard all too often. But even though Ansari's schooldays are long gone, that laughter, that mockery and abuse have found new avenues and forms through which to present themselves thanks to the ease of cyberbullying and, as of 2020, deepfakes.

As a filmmaker, Ansari has a public persona many queer people lack. And despite their religious background, Ansari's not afraid to be open about their sexuality. Not only that, Ansari's very work as a filmmaker does not shy away from that identity – it embraces it. Ansari's films explore and celebrate the queer distinctiveness they share with so many others. Yet it's Ansari's embrace of who they are in both their personal and professional life that's the reason they are still so often on the receiving end of abuse and bullying from the intolerant.

In February 2020, the trailer for Ansari's latest film, *Sheer Qorma*, debuted online. The film explores the relationship between two protagonists – one of them queer, Sitara (played by Swara Bhaskar), and one of them non-binary, Saira (played by Divya Dutta), both of whom also identify as practicing Muslims. The story is one of acceptance and identity and, for many, the trailer may have been their first exposure to the concept of people who identify as non-binary. In the trailer, Saira explains to a confused family member of their lover, Sitara, why Sitara refers to Saira as 'they'.

'I identify as non-binary. So my pronouns are gender-neutral. They, them, theirs,' Saira explains to the bewildered and unsympathetic family members. The intensity of the scene is powerful, and the remaining two minutes of the trailer shows the societal and familial struggles and conflict Saira and Sitara continue to encounter during the film due to the widespread stigma against homosexuality in India.

But that conflict and stigma were not limited to the film's story itself. Almost as soon as the trailer went live, Ansari tells me the hashtag #BoycottSheerQorma trended on Twitter. Yet, the calls to boycott the film were hardly the worst of the messages received. Soon, Ansari says they began getting an intense amount of hate on Twitter, Instagram and in their DMs. Those messages included threats of death, with many saying, 'we want to kill you', 'we want to murder you' and 'we want to throw you off the roof'. Others said, 'If you step out of your house, you want it.'

By August, the bullying and threats Ansari received had taken a new turn. One day they opened the Instagram app to find new DMs waiting. These messages featured more of the same textual abuse and homophobic slurs Ansari was sadly all too familiar with, but they also contained something new: videos of Ansari's face superimposed on scantily clad lingerie models and other sexualised women.

Their bodies. Ansari's face.

It was Ansari's first exposure to deepfake cyberbullying.

Ansari's tormentors created the deepfakes with ease. They used the Reface app, distinguishable by its watermark in the videos. The bullies merely selected source clips of scantily clad women and then uploaded an image of Ansari's face into the app, one taken from Ansari's Instagram account. Along with the deepfakes were homophobic smears and threats proclaiming 'watch this go viral' and 'watch how we are going to make you lose your dignity'.

The bullying I had experienced in high school suddenly felt tame in contrast to the videos showing up in Ansari's DMs. Words and fat jokes hurt, but this . . .

'Initially, I was ignoring it,' Ansari says. 'I was ignoring it, but after a point, it just gets to you, you know? It just comes to you, and it haunts you, and it gnaws at you, and it just makes you so goddamn uncomfortable.'

That gnawing discomfort took Ansari back to the traumatic experiences of the bullying experienced in their youth simply for being queer – that incessant laughter and mockery heard over and over and over again during their Bombay schooldays. But this was worse than historic sins.

'When things are done without your consent, especially to do with your face – because your face is really your identity . . .' Ansari says. 'That's really affected me in a very different way, in a way that I was not really ready for.'

Though we're speaking on the phone thousands of miles apart, I can hear the change in Ansari's voice. Something about it reminds me of old conversations at dormitory parties.

'Visually seeing yourself on a body that does not belong to you, created without your consent, then that being sent across social media – it really puts you in a spot that I really didn't ever imagine myself to be in. And it continues to be very traumatic.'

Given its impact, you might assume that when Ansari saw the first deepfakes sent by the cyberbullies, they might have been confused by what their eyes were viewing. Or at least confused by the process used to create the falsified clips. But

Ansari wasn't perplexed in any way. They're a filmmaker, after all – and a technically astute one.

Just the week before they received the abusive deepfakes, Ansari had actually downloaded the Reface app to try out themself. And, despite what's happened since, instead of shunning the app – the tool used to create their suffering – Ansari has embraced it. They still use the app and frequently post their own deepfakes of themself they've made with it. This might seem paradoxical at first, but by doing this Ansari says they've nullified their tormentor's weapons.

'It's about reclaiming your trauma; changing your trauma into a safe space,' Ansari explains. 'I think it's to do with the whole legacy of being a queer person. You know, it's like how the word "faggot" was historically used on queer persons, but then now, today, decades down the line, the community has actually reclaimed the word. The queer community has embraced it; they have accepted it. We now use the word "fag" as a badge of honour. And for me, to be honest, to still use the Reface app was truly about reclaiming my trauma. It was to go out there and put my own videos out there, which I see celebrate me, and to make sure that – you know what? You can keep doing the shit that you want, but I am now reclaiming that back. I don't give you the authority to do what you want. I can do it myself, and I can choose what I want to do with it.'

When Ansari tells me this, I can't help but smile. It's a response I wish my younger self could have been capable of formulating when I was bullied. What must it feel like to be able to grab your trauma by the collars and say, 'Let me show you just how much this doesn't bother me?' I know it's a strength my college friend would have wished he'd had in high school as well.

But I also can't help but wonder if Ansari's wisdom only comes with age. At 34, they're no longer that Bombay schoolkid. But what of others today who are that school age? How well will they be capable of handling deepfake bullying? Such bullying now embodies hardships that were unimaginable when I was in high school. Sure, I was bullied because of my body, but back then at least I had full control of it. Others could do nothing with my body itself – they couldn't use its likeness to harass and torment me. With deepfakes, that's no longer the case. Indeed, I think of a friend's daughter who told me about a heavyset girl at her school who has been digitally mocked by classmates using Reface – putting the girl's face onto the slim body of an R&B star shaking her booty in a clip from a music video. Will she be able to shrug the deepfake abuse off using Ansari's wisdom?

Some things only come with age.

I ask Ansari one other thing. Given the pain deepfakes have caused them – is causing others – do they lament the technology?

No, Ansari says.

'I don't really have a problem with technology, but I have a problem with the way we use technology. We are living in a world that is constantly booming; we are living in the age of globalisation; we are living in an age where people are virtually falling in love with one another despite having never met each other. So, there's nothing that we can really do about all this technology, which is enabling so many of these things. We cannot stop it, right? But how do you use technology? That still lies in our hands. The moral compass to use certain things still lies with us, you know? And we get to choose how we deal with it. When you have a child, you need to choose what kind of parent you're going to be, and it's the same with technology

– it's the same with deepfake apps, for that matter. You need to decide on how and what sort of moral understanding you are going to come to it with. And that really speaks greatly about who you are as a person.'

As someone who has both worked in the technology sector and then covered its societal impacts for a combined two decades now, I agree with Ansari's outlook. Technology – tools in general, as any technology is – doesn't have an essence. It's not good or bad. You can only pass moral judgement on the person who has used the technology in a certain way – not the technology itself. After all, a plane can be used as a tool to evacuate civilians from a natural disaster area, or it can be used as a tool to take down buildings. The plane is neutral in either situation. It is neither good nor bad. It has no objective. It's how its pilots choose to use that technology that our judgement should be reserved for.

Yet when it comes to deepfake technology, groups of people, far more powerful than Ansari and myself, are starting to disagree with our outlook.

The first? The Chinese Communist Party, which banned the Reface app in China last year. The reason? For 'disseminating politically harmful content'.[24] Though the Cyberspace Administration of China, the CCP's internet regulator and censor, didn't detail what 'politically harmful content' Reface was disseminating, many believe it had to do with the use of Reface by users outside of China's great firewall. Beyond the country's borders, Reface users were reportedly creating clips featuring Chinese President Xi Jinping's face deepfaked onto supermodel Miranda Kerr's body walking down the runway at a Victoria's Secret Fashion Show, and sharing them online.

Yet China's political leaders aren't alone in their willingness to take action against deepfake technology. The second group that wants to regulate the use of deepfake tech is none other than some US lawmakers. The reason, partly, owes not to fears of someone mocking the US president the way Ansari and Xi were mocked, but due to concerns over just what this rapidly evolving deepfake technology is capable of on other fronts.

Remember: back in 2014, when GANs were first created, they required hundreds or thousands of images of a target to pull off a convincing deepfake. By 2015, when the first smartphone deepfake app Face Swap Live came out, a face swap deepfake could be created in real-time via your phone's camera – though its results were still far from perfect, with some deepfaked faces looking like slushy snowmen. Yet by 2020 – just over four years later – deepfake tech in *smartphones* alone had advanced to such a degree that an app like Reface could make an ultra-realistic deepfake of someone based on a single selfie and place that person into Hollywood film clips with hardly a hiccup.

And if deepfake technology could evolve to do that in just four years, maybe all those headlines leading up to the 2020 elections weren't so alarmist after all? Indeed, while deepfakes didn't derail any US elections – yet – by the time of Ansari's bullying, a close cousin of deepfake – the shallowfake – was already successfully being weaponised against politicians and activists in the United States, causing havoc like never before. And their use, as we'll see, may be a very real sign of what is yet to come.

Chapter Two

Here Come the Shallowfakes

YouTube is the second most popular site on the web, only coming in behind Google's homepage.[25] It is accessed by three out of every four adults in the United States. Along with a monthly global audience of two billion, they go there to watch over one billion videos a day. And there's never a shortage of new content. On average, over 500 hours of new video is uploaded to the site *per minute*.

But most of those videos will hardly ever be watched by a large audience. In fact the average number of views most videos receive one year after uploading are estimated to be in the low four-figure range: 2,000 views; maybe 5,000 if the uploader is lucky.

So it's not surprising that professional content creators – people who make a living posting videos of everything from gadget reviews to beauty tips and pay close attention to You-Tube's algorithms, tailoring every upload to garner as many views as possible (from video length to thumbnail) – still get pretty excited when a video hits the 100,000 views mark. And if it makes it to half a million? *Outstanding results.* One million views? Break out the champagne, baby!

Needless to say, the champagne almost always remains corked.

So when a single video posted in 2016 by an unknown channel called Skitz4twenty[26] exploded to 3.5 million views – well, you can imagine it was a pretty big deal. Hell, most content creators I've known would probably sacrifice their grandmother, or at least a family pet, to have one of their videos hit 3.5 million accumulated views. 3.5 million views on a single video might as well be the moon for all but the biggest YouTubers. It's practically impossible to reach.

But for Skitz4twenty, it happened virtually by accident.

To call Skitz4twenty a 'YouTuber' or 'content creator' would seem laughable – probably both to him and to professional YouTubers (then again, their laughter would almost certainly be masking annoyed envy). Yet Skitz4twenty has had more success on YouTube than most people who post to the platform could ever dream of.

By the way, about that name, 'Skitz4twenty'. That's his YouTube handle, and he's asked me not to reveal his real identity. You'll soon understand why.

Skitz4twenty joined YouTube on 1 July 2016. A quick visit to his YouTube channel's homepage reveals how seriously he takes the platform. His profile pic is a small thumbnail of Lord Zedd, one of the main villains from *The Mighty Morphin Power Rangers*. His banner image – the branding billboard that all professional YouTuber's agonise over – is a simple YouTube logo on a black background with the letters 'FU' digitally spray-painted in lime green over the 'You'.

If that isn't enough to make most professional YouTubers aghast, the rest of Skitz4twenty's channel's page is sure to sicken them. The 'Channels' header has no other channel offerings (bad for SEO), the 'Community' header has no posts from the channel (horrible for keeping potential subscribers engaged), and the 'Playlist' header doesn't have a single playlist created for easy subscriber navigation (is he crazy?). Worst of all, under the 'Videos' header, there are only seventeen videos posted to the channel in total, the most recent one uploaded in April 2020.

Only seventeen videos in the channel's five years of existence.

All this is usually a death sentence for any YouTube channel. Yet Skitz4twenty's channel has a staggering 3,562,955 views as of the time of this writing. That from just seventeen videos by a YouTuber who clearly couldn't give a damn about the platform. What's even more impressive is those 3,562,955 views virtually all came from a single video Skitz4twenty posted.

A *single* video.

3,524,717 views (and counting).

That video was uploaded on 8 July 2016.

It is called 'The Hillary Song'.

The three minute and thirty-five-second video is a truncated clip from a longer televised event. Dwayne 'The Rock' Johnson sits on a stool in the middle of a wrestling ring in a packed auditorium. But The Rock's not in the ring to fight another match — that much is obvious. There's a microphone held in front of the wrestling and movie star's face by a boom arm. The Rock can't hold it himself because he needs his hands free for the guitar he's cradling. His biceps, each the size of two-gallon milk jugs, flinch and harden as he strums the instrument.

'There's a very special woman that The Rock wanted to sing a very special song to tonight,' he says as he strums the guitar and the crowd roars in the background.

It's a familiar chord.

'And I'd like to bring her out here,' The Rock adds as he continues to strum. 'Would you please come out? Don't be shy. Come on out. I just want to sing a song to you.'

After a few more strums of the guitar's strings, The Rock looks to his right. He smiles.

The person he wants to serenade has stepped onto the stage opposite the ring he's enclosed in.

It's Hillary Clinton.

As the video cuts from The Rock, we see Clinton approach a podium, US flags serving as the stage's backdrop, as The Rock continues his strumming and the auditorium roars at the sight of the then-presidential hopeful. Far from a wrestling match, this clip is a 2016 campaign event for Clinton – or at least a campaign stop at a wrestling match before the main event gets underway.

It's a stop that makes sense for Clinton. At the time, The Rock had never publicly endorsed a presidential candidate. He's also the rare celebrity adored by those on both the left and the right. Young males love his action movies, and those in the south and west of the country, especially, first became fans of him due to his days as a professional wrestler in the WWE. In other words, The Rock is respected and admired by a section of voters Clinton needs to sway to win the election. His endorsement could help put her in the White House.

'Ahhh,' The Rock says at the sight of her. 'There you are. You look so nice.'

As the crowd crows and The Rock continues to strum the same opening chords he has been, you can see Clinton mouth the words 'thank you' as she gives a coy smile.

Cutting back to The Rock, the film and wrestling star again says he has a very special song he wants to sing for her. He then breaks into a surprisingly beautiful rendition of Eric Clapton's 'Wonderful Tonight' (hey, some people have all the talents).

As The Rock serenades Clinton, the video intercuts between himself and the presidential hopeful. Clinton herself looks a bit uncomfortable at times, but she's being a good sport. And The Rock, he's bringing all the charm and charisma he can muster. He's not just singing the song from memory but tailoring some of the lyrics specifically towards Clinton, such as 'she brushes her short hair'.

The Rock continues among the trumpeting of the crowd.

'And then she asks me,' he croons, 'Do I look alright?'

There's a slight glint in his eye.

Another strum of the strings.

'And I say . . .' his eyebrows raise. 'No *beee-yatch* – you look *horrible* tonight.'

A devilish grin spreads across The Rock's face as the auditorium erupts with bellicose laughter at his upending of the classic Clapton song. And Clinton, as the shrieks and mocking chortling continues, she awkwardly looks down at her podium, appearing to mouth something.

Whatever it is, it can't be heard over the guffaws of the packed stadium.

As the clip then cuts to some in the crowd thunderclapping, The Rock commands the presidential hopeful not to go anywhere.

Cut back to a stunned, open-mouthed Clinton, who looks like she couldn't move if she wanted to. Even if you're not a fan of her, you can't help but feel for her. This is humiliating.

And for the next ninety seconds, The Rock continues his serenade of Clinton to the tune of 'Wonderful Tonight'. Yet, the words to the song are now completely bastardised by his ad hoc lyrics:

'You abuse all your power;
waste everybody's time.
You dress like a hooker;
not the expensive kind.
So get your ass to the airport;
take a one-way flight.
Because beee-yatch, you look horrible tonight.'

As the clip returns to a stunned Clinton, now wide-eyed, like a deer caught in headlights, The Rock has the packed auditorium join him in one last refrain:

'We said, "Beee-yatch, you look horrible tonight."'

The video ends with The Rock, his perfect pearly-white smile beaming, turning away from Clinton to address the ecstatic crowd, again lobbing an insult at Clinton's looks.

Skitz4twenty's uploaded video isn't long enough to see how the final scenes played out. The viewer doesn't know from the three minute and thirty-five-second clip alone if Clinton shot back at The Rock or if her politico handlers rushed her off the stage. But no matter which way it played out, any viewer of Skitz4twenty's video would agree: as far as presidential campaigning goes – this was *the* nightmare scenario. The insanely influential celebrity with almost unlimited star-power, who has tens of millions of voting-age fans on both sides of the political spectrum, and who's there *precisely* because he's supposed to drum up support for you, ends up roasting you in front of a packed auditorium's hysterically delighted crowd.

'The Hillary Song', as Skitz4twenty titled the clip, was only the third video he'd posted to his then week-old YouTube channel. The first video was a goofy 94-second clip of himself giving his cat 'catnip' and watching the results (currently 3,700 views to date). The second video: a 25-second clip of *Breaking Bad's* Walter White answering the door to find Mormons there (currently 395 views to date). But at the time of posting 'The Hillary Song', Skitz4twenty's first two videos had no more than a few dozen views each, and his channel had virtually no subscribers. There was no reason to think his latest upload, the truncated clip of The Rock's rude serenade, would reach greater heights. But then something happened.

In early August 2016, about a month after Skitz uploaded the Clinton-Rock clip, a friend contacted him with some news. He'd been browsing a Facebook group and saw 'The Hillary Song' reposted there. But not only that. Skitz's friend saw the video had been uploaded to a second Facebook group, too. It was being shared and liked. A lot.

That's when Skitz logged back into YouTube to check the video's stats. The last time he'd looked, 'The Hillary Song' had a few hundred views tops. But now? The views were in the hundreds *of thousands* . . . and rising rapidly. As the hits continued, Skitz's friend noted it's a shame Skitz hadn't watermarked the video with his name. That way, even though it was now being uploaded to other sites, people would still know who had created it.

Wait . . . *created* it?

Hadn't Skitz4twenty simply uploaded a truncated clip of a Clinton campaign event gone horribly wrong?

No.

What Skitz4twenty had actually done was create and upload what is now widely known as a 'shallowfake'. And his shallowfake, 'The Hillary Song', is still one of the longest and most viral examples of all time.

You see, the events portrayed in the Clinton–Rock clip Skitz4twenty uploaded to his YouTube channel never actually took place. Oh, sure – The Rock did really serenade-cum-roast someone to a bastardised version of 'Wonderful Tonight', and Clinton did actually walk out to a podium on a stage with US flags as the backdrop and mouth 'thank you' – just as Skitz-4twenty's video shows. It's just those two events weren't related.

They didn't happen at the same time, or even in the same place. Skitz4twenty just edited them together to make it look like they did.

While it's unknown which event the Clinton footage was taken from, it appears to be from one of her many speaking engagements from her time in office. As for the footage of The Rock strumming that guitar and serenading to the tune of 'Wonderful Tonight' – that actually happened, too. But it happened back in 2013, at a WWE Monday Night RAW event, long before Clinton announced her 2016 presidential bid. And far from serenading Clinton, the real person The Rock roasted was Vickie Guerrero, at the time, the general manager of World Wrestling Entertainment, Inc. (For those not familiar with professional wrestling antics, it's common for major wrestling stars to have public beefs with other wrestling personalities – it's all part of the show.)

What Skitz4twenty did was edit the footage of The Rock's serenade from the WWE event to remove all shots of Guerrero. Skitz then replaced the Guerrero shots with select shots of Clinton onstage at her event. Finally, by removing the original audio from the Clinton shots and keeping the original audio from the WWE shots even when cutting to the inserted Clinton clips, Skitz made the edited video seem like a recording of a single event.

The result was a video clip that looked authentic.

Well, to a degree.

'It's so awful,' Skitz tells me. 'It's cringe quality.'

He's being a little too hard on himself, but he has a point, which makes what happened next all the more baffling.

You see, when you watch Skitz4twenty's video, if you keep your eyes open, it's pretty apparent it's a fake. The intercutting between The Rock and Clinton isn't always clean. Sometimes Clinton appears to be silently speaking because the audio from her shots is removed, yet Skitz doesn't always cut back to The Rock in time before you see her lips move. There are even some obvious inadvertent jump cuts where Skitz had to shorten The Rock's dialogue to keep up the illusion that he was addressing Clinton and not Guerrero.

But Skitz wasn't ever trying to pass the video off as authentic. As you might have guessed from his utter lack of interest in maintaining a proper YouTube channel, Skitz made the video for a simple reason: he was bored.

'A friend downloaded Adobe Premiere on my computer, and I was just messing with it,' he explains over the phone when I call him to talk about the shallowfake. 'It was around election time, and every other ad or commercial or anything on TV was some election shit. So, I just started making videos – just kind of mocking them. And it was Vickie Guerrero's birthday, too, and so somebody had posted that Rock video clip online. And I was just like, "You know, wouldn't it be funny if . . ."'

So, Skitz edited the Clinton shots into the Rock-Guerrero roast.

'And I just kinda clipped it together and adjusted it, so it didn't have the WWE logo in it because I wasn't trying to get sued, you know?' he admits with a laid-back laugh. 'It was just for fun.'

Speaking of laid back, from my conversation with him, I think it's fair to say that Skitz is the type of person who, for me, wholly embodies that 'just for fun' lackadaisical attitude. I've interviewed everyone from tech titans to celebrities to famous novelists, all of whom are used to – or trained to – speaking with journalists. Yet, I've rarely been so quickly at ease with a subject as I am speaking to Skitz. He's the very definition of chill. I'm pretty sure most people who meet him would say he reminds them of that one relaxed guy we all knew in high school. You know the one. Though the thirty-seven-year-old lives in Phoenix, Arizona, there's a slightly southern drawl to his speech that makes you feel welcome. Like he's the kind of guy you'd want to sit and have a beer with on a porch on a hot summer night, just talking about life. His liberal use of the terms 'man' and 'dude' to address you – and 'dope!' to express his excitement – reminds me of where I grew up.

Something Skitz thought was dope, too, was that day in August 2016 when his friend excitedly contacted him, telling him 'The Hillary Song' was being shared in multiple Facebook groups.

'I was kind of geeking about it,' Skitz recalls. 'And I saw the views going up and everything, and I'm like, "Yo, this many people think this is funny?" You know, "That's awesome!"'

But then Skitz moved on from checking the view counter to reading the increasing number of comments accumulating on his video.

That's when the realisation hit him: 'Wait, these dumbshits think this is real?'

Yep. As Skitz so succinctly put it, the dumbshits on You-Tube and in Facebook groups thought 'The Hillary Song' was the recording of an event that actually happened.

Based on the comments on Facebook and YouTube, thousands — and maybe millions more, given the number of views and tens of thousands of 'likes' the video now has — thought Dwayne 'The Rock' Johnson duped Hillary Clinton, the former United States Senator and Secretary of State, into appearing at a wrestling match with the promise of being endorsed and serenaded by one of the biggest celebrities on the planet — only to be mocked, insulted, and ridiculed. And they thought that she — one of the most influential and accomplished people on Earth — would actually stay on that stage to endure that scorn until the guy with the neck the size of my waist was done humiliating her.

My first thought was: 'Dumbshits' is right.

While not everyone viewing 'The Hillary Song' thought it was real, it's evident from the comments just how many did. It's the WWE fans, like Skitz, who knew the context of the original clip that understood the obvious truth. However, many Trump supporters — and even some Clinton supporters — saw the clip and were either delighted or enraged.

Many expressed their joy that a megastar celebrity was willing to potentially risk his career by going against leftist Hollywood and not obediently supporting liberal politicians. The Rock, they said, wasn't afraid to speak truth to power, and they said their respect for him had only increased. Others asked why this event wasn't reported by the major cable news networks, implying a biased mainstream media. Still more simply congratulated The Rock for his ad hoc song, followed by calls to lock Hillary up — a popular rallying cry by Trump supporters. Some commenters even went so far as to suggest how Clinton could have handled the situation better.

And the comments went on and on.

'I was just like, "Holy crap. Why are you believing this? This is insane!"' Skitz says. 'Like, look at the quality. I mean, it's crap!'

But even some of the commenters who knew 'The Hillary Song' was fabricated thanked the video's creator for the edit, thinking Skitz4twenty was a MAGA supporter promoting a shared political agenda. Those being misinformed by the clip, they reasoned, at least were being misinformed in a way that diminished Clinton's profile and therefore boosted Trump's.

Yet watching some of Skitz's other videos posted to his YouTube channel, it seems obvious Skitz is no fan of Trump. One video Skitz uploaded is called 'Racist Donald gets trumped 3:16 style', which inserted some of Trump's most incendiary comments over a staged WWE confrontation with Stone Cold Steve Austin. Another video called 'Donald Trump Purge Ad' inserts horrific clips from *The Purge* movie franchise into an authentic Trump campaign ad, showing lunatics running around with masks and weapons pillaging and killing. (Fair play to Skitz – this video, it turns out, wasn't too far off from what happened during the attempted insurrection on the US Capitol by Trump supporters on 6 January 2021.)

Just to be clear, I ask Skitz directly if any of his videos had political aims.

'Politics in my eyes,' he says in that laid-back tone. 'If you're an asshole or something, I don't like you. You know, Trump should have never been taken serious running for president. Like that guy is a moron. Hillary . . . I just didn't like her either. I went third-party on that one. Cause I'm like: neither of you people deserve a vote.'

And so, misinformation of any kind wasn't part of Skitz's plan for 'The Hillary Song'?

'No. No misinformation – nothing,' he insists. 'Like I was just bullshitting pretty much. It was me having fun with the new program, and I didn't really know how to use it. I was just having fun.'

And that, right there, is what really brought it home for me. It worried the hell out of me – it should worry the hell out of you, too.

I believe Skitz when he tells me he was just having fun – just trying out a new basic video-editing app a friend loaded onto his computer. I believe that Skitz, out of sheer boredom, edited a video and uploaded it to his YouTube channel. But from there, that crudely edited video – that shallowfake – was shared on one Facebook group, then on another, and then shared even more widely on other online forums – most viewers believing the clip to be real. And in less than a month, his video went from zero views to hundreds of thousands, to today, where it has over 3.5 million views. Never mind that Skitz never meant for it to be used and viewed as a political statement. Never mind that the quality was . . . well, pretty poor. It went viral – and not only did millions of people watch and share it, many *believed* it too. They believed it, even though there's not really any reason they should have.

How accurate that old saying is: 'Seeing is believing.' But I'm getting ahead of myself.

The thing about shallowfakes is that, just like deepfakes, they create a narrative that is ultimately misleading; they show something that, in reality, did not happen. But unlike a deepfake, which relies on deep machine learning techniques, a shallowfake relies on traditional editing tools and manual labour to alter existing media.

In fact, they are often referred to as 'cheapfakes' because, due to the manual labour and human editing skills required, some can come off looking cheaply made if the human creator doesn't have the necessary expertise in graphics and video editing.

Skitz's shallowfake may have been one of the first such pieces of this type of media to go truly viral, but they've been around for much longer than a few years. Before their colloquial name was derived from the genius younger sibling that is deepfake technology, shallowfakes were known by a different label – 'manipulated media'. And 'The Hillary Song' was far from the first manipulated media.

For as long as there has been photorealistic visual media, people have been manipulating it. Airbrushing, anyone? That manipulative technique goes back to at least 1876, when it was first patented. Cropping? That was first used on photographs decades earlier. What about gel filters over the lens on a film camera to make a shot appear taken at dusk instead of high noon? Such film manipulations are almost as old as the industry itself.

What is unique to shallowfakes is the ease with which they can be both created and spread because the tools and means to do so are now ubiquitous in the digital age. Unlike deepfakes, which still rely on obtaining specialist software, your smartphone or laptop comes pre-installed with all the software you would ever need to create a shallowfake. Combine this software with a nearly unlimited supply of videos of individuals – famous or not – already posted to the web, and you've got everything you need to make a shallowfake on your computer this very instant.

And thanks to the internet and social media, these manually altered shallowfakes can spread around the world at a startling pace. Just look how quickly Skitz's video went from a few dozen views to a few million. But while Skitz manipulated his selected media out of boredom and he uploaded it with the hope that – at most – a few people would get some harmless chuckles out of it, the weaponisation of shallowfakes by people with more nefarious objectives took off in the years immediately after 'The Hillary Song' went viral.

On 14 February 2018, a man walked into Marjory Stoneman Douglas High School in Parkland, Florida, and activated the fire alarm. When students and teachers evacuated their classrooms, the man opened fire. In under six-and-a-half minutes, the gunman killed fourteen students and three staff members. Another seventeen people were injured.

In the days that followed, the response, on a national level, was all too predictable. Politicians in the pockets of the powerful pro-gun lobby group, the National Rifle Association, offered their 'thoughts and prayers' to the victims and their families. There were the familiar platitudes but no serious talk about meaningful change. 'Now's not the time.' 'It's too soon,' the script always goes. 'Guns aren't the problem; unarmed teachers are.'

As a journalist who has written about the country's unending string of mass shootings too many times, I can tell you how maddening these pre-packaged, empty comments leave you feeling. You can predict the exact words, and the people who will spout them, before they've even parted their lips. It is always the same – because when it comes to gun violence in the United States, nothing ever changes. Sadly, what comes in the days following the bloodshed and rote comments is just as predictable.

The Parkland school shooting, as all US school shootings end up becoming, was quickly politicised. It became not about the tragedy of young people being gunned down en masse, but instead twisted into an intentionally divisive narrative about those on the left wanting to take away Americans' Constitutionally guaranteed right to bear arms. And, of course, as if the whole tragic situation was just an over-the-top satire, gun sales in Florida actually *increased* after the shooting. All this while parents buried their slain children, as young as fourteen.

But one group wasn't having it any more. That was the victims themselves – the survivors of the Parkland massacre. On 17 February 2018, several of the Marjory Stoneman Douglas High School students and their friends and family held a rally outside of the Broward County Courthouse in Fort Lauderdale, Florida, to call out the politicians who, school shooting after school shooting, fail to act to protect the country's young.

One of those survivors was eighteen-year-old Emma González, who gave a powerful eleven-minute speech that would launch her activism career.

'Politicians who sit in their gilded House and Senate seats funded by the NRA, telling us nothing could have ever been done to prevent this – we call B.S.!' she called out to the crowd.[27] 'They say that tougher gun laws do not decrease gun violence – we call B.S.! . . . They say that no laws could have been able to prevent the hundreds of senseless tragedies that have occurred – we call B.S.! That us kids don't know what we're talking about, that we're too young to understand how the government works – we call B.S.!'

The speech quickly went viral, and within weeks González became a poster girl for the much-needed gun control in the United States. By 23 March, *Teen Vogue* published[28] a photo-shoot of, and op-ed by, González in which she attacked the NRA and called out the ludicrousness of those saying school shootings could be prevented if we merely armed teachers with guns themselves. And on 24 March, she, along with other Parkland survivors, attended the March for Our Lives rally in Washington, DC, where she gave another highly praised speech in front of hundreds of thousands of supporters demanding stricter gun control laws. This speech lasted six minutes and twenty-seconds – the length of time the Parkland massacre took place in.

It was just after the March for Our Lives rally ended that González found herself in the crosshairs again. Although this time, the weapon targeting González was not a gun: it was a shallowfake.

An animated GIF was being shared widely on social media. In it, the eighteen-year-old activist, shaved head and wearing a tight black top, held a copy of the US Constitution in her hands – one of the primary symbols of the United States, and one deeply important to gun rights activists due to its Second Amendment guaranteeing the right to bear arms. The González in the GIF then rips the Constitution in two before discarding the halves and folding her arms over her chest in a defiant manner.[29]

The González in the GIF is very real, of course. It is actually her face, her body and even her actions. And she really was tearing a large piece of paper in half in the original video the five-second GIF was created from. However, the paper she tore in half in the original video wasn't the US Constitution. It was a large poster with a bullseye printed on it – the kind used as shooting targets at gun ranges. González filmed the original video as part of her *Teen Vogue* photoshoot days prior, and it was only first posted online twenty-four hours before the March for Our Lives rally. What someone angry about

43

González's gun control activism did during that brief timespan was take that *Teen Vogue* video, use digital editing software to manually edit the bullseye out of the target González held, and then digitally insert the words of the US Constitution in its place to misrepresent her actions.

It was character assassination.

And it was a good shallowfake. It looked real because whoever made it clearly had a certain mastery over the digital editing application they used to create it. And it was a shallowfake that was shared as fact hundreds of thousands of times across pro-gun and right-wing social media channels before fact-checkers could verify its inauthenticity the following day. Yet by that time, the shallowfake had achieved its aim: discrediting González in many people's eyes by showing her doing something that is seen as profoundly insulting to most Americans – whether you support stricter gun control laws or not.

This was just the first in a long line of politically motivated shallowfakes beginning in 2018 and leading up to the contentious 2020 elections. And from González's 2018 shallowfake, the deployment of shallowfakes as political weapons only accelerated until even the president of the United States himself was distributing the manipulated media online.

The next major politically motivated shallowfake came a little over a year later. But this time the shallowfake's aim was not to discredit an activist, but the president's most formidable political opponent.

In May 2019, a shallowfake of House Speaker Nancy Pelosi, the highest-ranking Democrat in government at the time, hit the web.[30] Like the González shallowfake before it, this one was designed to impugn the target's character. The thirty-second shallowfake shows Pelosi speaking at a Center for American

Progress event. However, her speech is sluggish and slurred as if she's drunk. Pelosi did not sound like this at the original event. Instead, video of that event was slowed by 75% of its original speed to help achieve the desired effect. Yet that's not the only alteration the shallowfaker made.

Typically, slowing a video by three-quarters its normal speed would affect its audio to such a degree that it would be evident to the viewer that the voices in it were speaking in slow motion, and thus give the ruse away. To get around this obvious tell, the creator of the shallowfake used additional audio editing techniques to change the pitch of Pelosi's deeper, sloweddown voice to make the protracted audio sound more like her natural speech – well, an inebriated Pelosi's speech. These sinister alterations, taken as a whole, made it look and sound as if the highest-ranking Democrat in government had been plastered at the event. While Trump did not tweet this particular shallowfake, numerous right-wing accounts and even the president's personal lawyer, Rudolph Giuliani, tweeted a link to it before eventually deleting his tweet.[31]

Once the ball got rolling, the shallowfakes just kept coming. And with the combined millions of shares these doctored videos garnered – and the damage they caused – is it any wonder that the weaponisation and deployment of shallowfakes quickly went into overdrive once election year arrived?

Barely two days into 2020, a shallowfake of Joe Biden made its way into millions of Americans' social media feeds.[32] The shallowfake was edited to take a part of a speech the presidentialhopeful made days before out of context. In the shallowfake, Biden says, 'Our culture is not imported from some African nation or some Asian nation.' Critics leapt on this, accusing the Democrat – the party of a majority of African Americans – of racism. But listening to that part of his speech in its full context revealed Biden was speaking about how violence against women was 'not imported from some African nation or some Asian nation.' Rather, he said such violence had its origins in 'our English jurisprudential culture, our European culture.'

If that doesn't sound nearly as high-tech as the 'drunk Pelosi' video, it's because it's not – manipulated media can be as simple as shortening a clip. You can do that in thirty seconds on your laptop – hell, you can do it on your smartphone. Taking quotes out of context is not something that is new in media at all; it's a tale as old as time, a tale at least as old as propaganda itself. But manipulating the context of media at the rate that occurred in 2020; purposefully misleading viewers – and doing so in a way that for some reason they – we – still implicitly trust . . . that's the road we're on now. And there's certainly no turning back. After all, each shallowfake is just another stop on the highway to deepfakes, at which point these media manipulator masters will switch over from the relatively archaic bus they're on to the high-tech bullet train and leave our heads spinning in the process.

As for 2020, there were still plenty of stops for the bus to make. Nancy Pelosi's turn soon came again, this time in the form of a new shallowfake of the House Speaker which was shared on Twitter by President Trump himself.[33] If you thought the views Skitz's shallowfake got were impressive, imagine the engagement a shallowfake shared by the President of the United States garnered – that's to say nothing of the fact that we even need to wrap our heads around the fact that a sitting US president is sharing shallowfakes in the first place. This shallowfake showed Pelosi ripping up Trump's 2020 State of the Union speech just as the president introduced and saluted one of the last living Tuskegee Airmen from World War II – the ultimate dis-respect. Granted, Pelosi *had* torn up the speech – but she did so a full fifty minutes later. In fact, Pelosi had stood and applauded the Tuskegee Airman when Trump introduced him.

As for who created it? No one knows. But the identity of the shallowfaker is irrelevant next to the ability of the creator's manufactured narrative to spread far and wide. And that spread – well, we're to blame. Take an overly polarised population, throw in social media at the height of its power, and add in the capability to manipulate what people see with their own eyes – to control the narrative – and the only surprising (and surprisingly consistent) part of all this is how inept the general public has remained at spotting a shallowfake. We are about as bad at mistrusting falsified media as we are good at sharing it.

Next up was Biden, again, who was now the clear frontrun-ner to take on Trump in the 2020 election. This shallowfake, released in March 2020, featured a video edited to make it look like Biden was endorsing Trump, saying, 'We can only re-elect Donald Trump.'[34] The White House social media director tweeted the shallowfake, which was then retweeted by Trump himself – of course reaching millions within minutes. In reality, the shallowfake truncated the original video in which Bid-en explained that Democrats needed to take the high road in the presidential campaign. This shallowfake has the distinction of being the first media ever tweeted by a US president that earned Twitter's new 'manipulated media' label to let viewers know the clip they were watching was inauthentic. Of course, that label is a bit like putting a bandage on a gunshot wound.

And just a few months later, in June 2020, politically motivated shallowfakes hit a new low. This time the sub-jects of the shallowfake were two toddlers. The shallowfake purported to be a CNN news story that showed a Black tod-dler running away from a white toddler. Under the doctored CNN logo, a Chiron read, 'Racist Baby Probably a Trump Voter'.[35] The original video, in fact, showed the two tod-dlers hugging before the Black one ran off. However, the brilliance of this shallowfake was not that the creator was trying to make the audience believe a white toddler was be-ing racist against a terrified Black toddler, but rather that was what CNN was reporting to its audience. In other words, the shallowfaker wanted to trick viewers into thinking CNN broadcast a ludicrous and obviously fake story. The intended narrative? Racism in America is not the problem – fake news alleging racism in the United States is. How very meta. This shallowfake, too, was shared on Twitter by Trump.

And then came the final days of August 2020 – and the use of shallowfakes went off the charts. With just over nine weeks until election day – and Biden leading Trump in the polls – some of those on the political right pulled out all the stops.

On 30 August, the White House social media director shared a new shallowfake.[36] In it, Biden appeared to be asleep during a live interview. This shallowfake was created in much the same way Skitz created 'The Hillary Song': by editing two separate events together to make it look like one. In the original Biden footage, the presidential candi-date briefly glanced down before looking back up at the camera. Yet this brief action was edited and combined with footage from an unrelated 2011 Harry Belafonte interview in which the actor did appear as if he had fallen asleep, and the remote interviewer had actually awkwardly urged him to 'wake up' (Belafonte said the original incident was due to a technical glitch[37]). But in the shallowfake, Belafonte was edited out and replaced by the shot of Biden glancing down. Added to this edit was a loud snoring sound effect, which appeared in neither of the original clips.

Also, on 30 August, the number two Republican in the House, the Minority Whip Steve Scalise, shared a different shallowfake.[38] This one altered the conversation between Biden and activist Ady Barkan, who has ALS (amyotrophic lateral sclerosis) and speaks using a computerised artificial voice. In the shallowfake, Barkan's computerised voice was altered to make it sound like he asked Biden if, as president, he would 'redirect some of the funding for police' to which Biden answers, 'Yes'. In reality, Barkan asked Biden if he would redirect some law enforcement funding towards mental health services, which Biden said he would do. The 'for police' spoken by Barkan's computerised voice in the shallowfake was spliced together from a different part of his conversation with Biden.

In response to Scalise sharing the shallowfake, Barkan tweeted,[39] 'These are not my words. I have lost my ability to speak, but not my agency or my thoughts. You and your team have doctored my words for your own political gain. Please remove this video immediately. You owe the entire disability community an apology.' Twitter labelled the video 'manipulated media' and Scalise ended up deleting it.

You may think that abhorrent shallowfake would make even the most egregious offenders scale back their manipulation campaigns. But no. On 31 August, the Trump campaign's 'War Room' Twitter account tweeted[40] a three-second clip of Biden proclaiming, 'You won't be safe in Joe Biden's America.' Another shallowfake that took the original quote out of context. You see the pattern here.

And just days before the presidential election, on 1 November, a new shallowfake was shared widely on social media. This shallowfake appeared to confirm the narrative Republican operatives had been building around Biden throughout 2020 – that he was mentally unfit to serve as president (the irony). In the shallowfake, Biden appears to be addressing a crowd in Florida but embarrassingly calls out, 'Hello, Minnesota!'[41]

No points for guessing where Biden actually was when he said that. The shallowfake was edited to make signs in the footage read Florida.

I'd love to tell you that shallowfakes were simply part of a nasty election year. But I'm sure it won't shock you to hear that they're here to stay, and these instances were just the beginning. Today many of these shallowfakes and others are widely available as GIFs on popular image-sharing websites, meaning they can easily be propagated around the web and in private messages by those who wish to sow disinformation – or those merely believing the shallowfakes are real.

Speaking of believing shallowfakes to be real, there is one unexpected addendum to Skitz's 'The Hillary Song'. As has become somewhat of a theme with shallowfakes, regardless of quality, a large majority of the commenters clearly believed 'The Hillary Song', too, to be the record of an actual event. And that's why, if you are one of these people, you could be forgiven for being baffled on 27 September, barely five weeks before the 2020 presidential election. On that day, The Rock shared a video on Twitter.[42] But there was nothing fake about it – shallow or otherwise.

In the video, The Rock conducted a remote interview with Joe Biden and Kamala Harris discussing their goals for the country should they defeat Donald Trump in November.

Along with the video came a shocker: 'As a political independent & centrist, I've voted for both parties in the past,' The Rock's tweet read. 'In this critical presidential election, I'm endorsing @JoeBiden & @KamalaHarris. Progress takes courage, humanity, empathy, strength, KINDNESS & RESPECT. We must ALL VOTE.'

After decades of remaining mum, The Rock had finally, for the first time in his life, publicly endorsed a candidate for President of the United States – and a Democrat, no less. And, well, you could see why many of his right-wing fans who had vicariously lived through The Rock's hypermasculinity onscreen and in the ring for so long would be upset – perhaps even feel betrayed. Their comments on his tweet certainly reflected betrayal.

Calls of 'traitor' and 'fool' were commonplace. Others suggested his endorsement of Biden meant he was anti-police, pro-socialist, and pro-China. Some made baseless claims The Rock only endorsed Biden because he was being blackmailed, perhaps by Hollywood or the Deep State boogeymen. Others announced they will never watch a Rock movie again and that his career in the film industry was over because his fanbase would now desert him in droves.

But those weren't the only types of comments left.

Other commenters responded with links to Skitz's 'The Hillary Song' posted on YouTube over four years earlier. Along with the link, they asked The Rock why he flipped from hating Democrats to supporting them? How could he go from doing what he did to Clinton in 2016 to endorsing Biden in 2020?

Go have a look for yourself. Even checking the most recent YouTube comments on Skitz's work shows that many still think the shallowfake is authentic a full five years after it was posted. Many tried to wrap their head around why The Rock endorsed a Democrat this time around, hurling insults at both him and Biden. They asked why he changed so much. They called him a hypocrite and a back-peddler. A motherfucker. A baby-eater. A sell-out. But most of all, many couldn't seem to get past the cognitive dissonance that emerged from seeing The Rock very clearly make a fool of Clinton four years earlier yet now endorsing her political ally and ideological successor.

The lingering power of altered media is astonishing. The hold it has over people – even five years later – is unreal.

As a matter of fact, it was this sudden renewed increase in comments on 'The Hillary Song' after The Rock endorsed Biden that drew Skitz back to his old shallowfake, which he hadn't thought of in a long time.

'I started seeing the comments: "Traitor!", "I'm never watching another Rock movie again,"' Skitz says in disbelief, 'and I'm like, "Oh! You people *still* believe this?" Like even after all the misinformation that's out there, you don't question this? *Still*?'

I ask Skitz if the power these shallowfakes – shallowfakes like his – have to mislead people even after all this time . . . does that unnerve him? After all, his shallowfake was one of the first modern ones to go viral, and in the time since he created it, they've only become more frequent, and more weaponised. With the ease at which they can be made, just as he did his using a basic app on his computer – does he think things will get better or worse?

'It might get worse just because people are getting more skilled with it,' Skitz admits, noting how, in the past year especially, people have been sitting at home, picking up new skills out of boredom – anything to pass the time as the virus rages. It was a similar boredom and access to free software that led him to create 'The Hillary Song' in the first place. 'And I had no skill whatsoever, you know?'

Yet look at the sway his unskilled shallowfake still has over so many people all these years later.

And Skitz and I didn't even get into deepfakes – where you need virtually no artistic or creative skill to produce fake videos that are an order of magnitude more realistic-looking than even the best shallowfakes are now. And not just more realistic looking. Deepfakes allows anyone to manufacture wildly more complex falsified scenarios with ease. You no longer need to pore through real footage to find a single line of spoken audio you can edit out of context for your shallowfake. You can simply deepfake that target to say – or do – anything you want them to – and with fewer manual skills than even Skitz needed to learn to use Adobe Premiere.

That fact itself helps explain many of those seemingly alarmist headlines about deepfakes' threat to the election in the run-up to the 2020 vote that I mentioned earlier. Now can you see why – if a glut of politically weaponised shallowfakes were causing all this trouble in the last few years – why so many were worried that not them, but the right weaponised deepfake would have no problem altering the outcome of the 2020 election?

Though shallowfakes proliferated throughout 2020, in a way, we were lucky it was them – because their very nature makes them self-limiting. Shallowfakes do not improve on their own. Even if a shallowfake creator swaps their pre-installed iMovie app for professional Hollywood video editing software like Avid Media Composer, the shallowfakes themselves don't automatically get better. Any increase in quality depends on human creators becoming more skilled at manual editing and graphics techniques using whatever software tools they can access. This requires training and time, which limits the production of shallowfakes, and, ultimately, their spread. And as even a gifted shallowfake creator needs to eat and sleep like the rest of us, there is only so much convincing output they can generate in a day, week, or month.

The same cannot be said for deepfakes, which only rely on a human insofar as much as the deepfake software needs to know what the human's objective is: what the video to be altered is and who's the target to be inserted? But once the human provides those two variables, the software itself fabricates the desired deepfake without much additional human effort. And its AI gets exponentially better the more it creates – and it gets better far more rapidly than a human using the manual techniques that shallowfakes rely on ever could – even if that human had multiple lifetimes to hone their skills.

You might say that shallowfakes walked so that deepfakes could run. But while no shallowfake derailed the election, their weaponisation did cause real-world harm.

Shallowfakes aim to generate anger, outrage, indignation and feelings of self-righteousness in the viewer. These emotional states only further entrench people into their political tribes – causing them to view the world through an 'us versus them' mentality. This mentality keeps us divided from one another. And not only that.

Since we, the flawed beings we are, often cast aside reason and rational thinking when we are whipped up into a state of anger and outrage, it's easy to understand how one-too-many shallowfakes may contribute towards us physically turning on one another – you simply have to view footage of the 6 January 2021 insurrection and attack on the US Capitol to know what I'm talking about.

And while shallowfakes didn't directly lead to that attack, they were part of a set of sustained and constant lies and misinformation fed to a group of people desperately needing to believe in such a narrative in order to feel like they had understanding over a complex world that so often left them feeling dissatisfied and ignored. This understanding, while false, and in part propagated by shallowfakers and those who shared their work, nevertheless gave the group a roadmap, no matter how skewed, that directed them where to go – and hinted at how to act. In the process, it made some of them feel like they had agency – however misdirected – over that unique point in US history.

And we've seen how that turned out.

Speaking of how things turned out . . .

Dwayne 'The Rock' Johnson wasn't the only one to vote for Biden in 2020. Viewers of 'The Hillary Song' may be surprised to learn that Skitz voted for Biden as well. That's one reason he didn't want me to reveal his real identity in this book.

'I don't know how badly people come after people on the internet just for stupid shit. But obviously, if they believe *this* . . . you know?' Skitz says, though still as chill as ever. 'I'm thick-skinned, but I don't want them actually doing something stupid.'

I can't say I blame him for the abundance of caution.

I also can't solely blame shallowfakes for the seeming hysteria about the impact their more capable and realistic cousin, deepfakes, could have had on the 2020 elections – and society at large.

That's because the worries about deepfakes were not solely based on hypotheticals: 'If shallowfakes can do this, *imagine* what deepfakes might be able to do one day.'

There was no imagining needed.

You see, in the years leading up to the 2020 elections, it wasn't just shallowfakes being weaponised. Deepfakes were already being weaponised, too – and I'm not just talking about bullies using face swap apps to harass and intimidate their victims, as horrible as that is.

Since 2017, while ever more versatile deepfake-powered face swap apps and funny fancast YouTube videos were entertaining much of the world, in forums and in private chat rooms across the internet, more advanced deepfaking software was being regularly and systematically used and weaponised against a specific group of people. A group much larger than all the politicians and activists in the world combined: women.

Chapter Three

Never Is a Lot Shorter than It Used To Be

I still remember the first time I saw Sandra Bullock naked. It was a spring night in 1999, and I had just returned to my dormitory after having been out to see a movie. I walked up the first flight of stairs in the lobby and said hello to the night security guard seated in his usual place behind the desk. We chatted for a few minutes after I pressed the elevator call button. It was taking longer than usual.

When it finally arrived, I said goodnight to the security guard and stepped inside the elevator. I pressed the button for my floor and waited for the door to slide shut, which it always did slowly and with a scrape. As I watched it creep across my vision from its recess, instead of being greeted with its metallic sheen, a sheet of computer paper taped to the inside of the door slowly revealed itself as if it were walking onto a stage. The paper was taped at all four corners, the scotch tape extra-wide to ensure it wasn't going anywhere. On it was printed a picture of Sandra Bullock, who at that time, in the late 90s, was one of the most popular and in-demand actresses in Hollywood.

There was her signature brunette hair, her beaming smile, and her prominent cheekbones. From the neck up, it was the perfect PR shot. From the neck up, she looked as if she had just stepped onto the red carpet. But the rest of her . . . well, there is no polite way of saying this. There was not a speck of clothing on her. She was completely nude except for, oddly, a thick athletic wristband on each wrist.

The thing is, I know major Hollywood stars have the money to hire personal dieticians and trainers to stay in peak physical shape – but Bullock's naked body was . . . it's hard to describe it, but it was almost *too* perfect. There wasn't an ounce of fat anywhere, every inch of her was perfectly toned or curved, and her skin was uniformly tanned – not too little or too much. There wasn't a blemish to be found.

I don't know if you've ever seen anything that's just too good to be real – so much so that it's suspicious in its flawlessness. But this was the picture that had appeared on the inside of those elevator doors. And the more I looked, the more perfect – and uncanny – it seemed. Yet there was one fault – though I suppose many wouldn't call it a fault at all – that jumped out at me. It was with the breasts. Don't get me wrong – they, as the rest of the photo, were flawless, as far as breasts go. In fact, they looked like the ideal – if you polled every person with an interest in breasts, they would probably describe something close to the picture in front of me. The thing is, they were a little . . . too big? I don't mean cartoonishly so; not Jessica Rabbit-too-big. I mean, too big for Sandra Bullock. From every movie I had seen Bullock in, she always appeared to have a slim build. Yet in this photo, her breasts were anything but. They simply didn't match the Sandra from the films I was familiar with.

As I stood there, perplexed, trying to reconcile this discrepancy, those too-large breasts slid away from me, along with the elevator's door as it retreated into its recess. I'd arrived on my floor.

But I didn't move.

I stayed in the elevator until its door slid shut again, and rode that elevator up to another floor, where it stopped, only this time I ripped the image from the door before it opened. The only proof something had ever been there were three of the four torn corners that remained scotch-tapped in place.

When I finally entered my room, one of my friends was watching TV. I held up the torn image of Bullock, and my friend turned his head over his shoulder to look at it. He gave a small laugh.

'Is this real?' I said.

Instead of answering, he pulled a folded paper from his pocket. Another Bullock. The same picture.

'Someone put a bunch of them up all around the dorms,' he said.

'But it's not real, right?' I replied.

'What do you think?'

I shrugged.

'It's Photoshop,' he said.

Now, even though this was the late 90s and Adobe Photoshop had been around for almost nine years, your average person still had probably not heard of the program. Photoshop and the verb 'Photoshopped' really didn't become commonly known until the early 2000s when the first social media sites started taking off. But back in 1999, both I and my friend – and almost everyone in the dorms, for that matter – were well acquainted with Photoshop. You see, I went to a film and arts college, and as a result, almost everyone in the dorms was a film, graphics, photography, or other media major of some kind. Not only did virtually every student living in that dorm know about Photoshop, but many of us also had it and other editing applications on our computers. In other words, while students at most colleges thought Word and PowerPoint were cutting-edge programs and they might be able to turn out a hell of a slick-looking report using them, most at my college could cut you a twenty-minute sizzle reel with titles and graphics that rivalled at least mid-tier post-production houses.

'Who made it?' I asked.

My friend answered with a shrug.

I examined my copy of the photo again. Bullock's beaming smile. 'The quality's amazing.'

'Breasts are too big,' he said.

'Yeah, but if you didn't know who Bullock was . . .' I remember saying. 'I mean, the way the face blends onto the body at the neck . . . it looks real.'

What I meant by that was, essentially you couldn't tell this was Photoshopped. The only 'tell' to it being so was that I happened to know better because the creator of this fake image had used an actress whose build most of America's movie-going public was familiar with. But the technical proficiency of the person behind this Photoshop creation was off the charts. The skin tone of Bullock's face and neck blended perfectly to match the skin tone of 'not-Bullock's' body. Even the lighting adjustments were incredible – the shadows around her nose matched the direction of the shadows cast on the rest of 'her' body. Besides the breasts being too big and the weird exercise wristbands, this fake photo was nearly perfect – and by perfect, I mean entirely realistic. It must have taken its creator dozens of hours – maybe days – to do.

'And you don't know who did it?' I tried again.

'Nope,' my friend said, turning back to the TV.

This was my first exposure to fake celebrity porn – but throughout my final years of college, these fake celebrity photos occasionally appeared – taped to the back of an elevator door or in one of the stairwells. I never found out who created them, but soon their appearances spread outside the dorm, too. In classes, you sometimes heard students mention them in whispers – students who didn't live in my dorms.

'Did you see that naked photo of Jessica Alba?' someone in my film editing class asked.

'No, is she doing topless scenes now?' the other replied.

'No, dude. It's fake,' the first explained, saying that another classmate who liked Alba found it on an online forum. A forum where people who use Photoshop to create fake nudes of celebrities share their work.

This was the first I'd realised that fake celebrity porn was bigger than just something being spread around one film school. This realisation moved me beyond any naive idea that – at most – a few geeks with computing graphics skills were doing things just because they could, or because they were bored. There was an entire active internet subculture of fake celebrity pornographers – and it was growing rapidly along with the demand for this new kind of Photoshopped content.

Of course, it's fitting that I first found out about fake celebrity porn at my college. We were all film buffs, after all, and going to film school because we were so enamoured with the industry – an industry that is by its very nature artificial and glamorised; where people are products, and the products are worth $200 million opening weekends if you put them in the right films with the right co-stars. But in a way, even we in film school weren't that different than most in the United States. As a whole, we're a society that has always been obsessed with celebrity – and sex. We invented modern celebrity culture – it was built right into the Hollywood star system born in the

1920s. 'Sex sells' in advertising? That was the United States, too. And we exported both constructs across the globe. By the mid-90s, the average American probably knew more about who Gwyneth Paltrow was dating than the fact that there was a recent genocide in Rwanda. *Keeping Up with the Kardashians*? Yes, please, let's keep up. A Hollywood sex tape scandal? All the details are coming at 6.15, right after the weather – stay tuned.

Considering all this, it wasn't a stretch to recognise the appeal of fake celebrity porn. After all, Hollywood already sells us pre-packaged scripted fantasies starring an ever-revolving door of young actresses: we see them in danger, in love, sexualised, and opposite them on screen, there is always – always – some hunky (yet relatable) love interest. Hollywood loves leading men playing characters that average men can project themselves onto; it feeds into the whole Hollywood fantasy. Have you ever heard the phrase 'women want to be her; men want to be *with* her'? This is essentially the promise that Hollywood has spent decades selling.

And what fake celebrity porn does is take that Hollywood promise one step further. It continues playing the silver screen scene that would typically fade to black: that scene where the guy gets the girl, and they each get what comes next in any romantic relationship. Fake celebrity porn is, in many ways, the natural next step in the fantasy that Hollywood flogs.

It was this absurdity of not just fake celebrity pornography, but America's obsession with celebrity, and Hollywood's long history of its willingness to turn people – flesh and blood people – into packaged commodities, sometimes with deadly effects (just look at what happened to Judy Garland[43] – one of many), that inspired me to write my first novel, *Epiphany Jones*. It's a novel I wrote after becoming disillusioned with the whole industry.

The story follows Jerry Dresden, a man addicted to fake celebrity porn, as he's thrust into an unwilling relationship with a woman entangled with a sex trafficking ring that caters to the Hollywood elite. In addition to exploring the underbelly of the glamorised facade Hollywood would like us to believe was real, the novel confronts the way we — the voracious consumers of Hollywood content — look at celebrities as little more than packaged products, too, and it examines the internet subculture made of those who spend their free time creating fake celebrity porn, allowing those who want to take the Hollywood fantasy one step further to do so.

While the novel is, of course, fiction — the subculture of those on the internet who create Photoshopped fake celebrity porn is very real. And while it might be easy to simply dismiss the people creating these images — the 'fakirs', as they are known — as perverted, or opportunistic, some of them have motives you wouldn't expect. You see, I spent quite a bit of time tracking these people down during my research for the novel. I wanted to know their reasons for creating the content they create, to understand more about *why* they did it.

Some, of course, did just want to see their Hollywood crush engaged in highly graphic sex acts — they wanted the snapshot in their mind to be the image on their screen. Other fakirs, as you might expect, did it for money. It can be a pretty lucrative market, after all.

I've also met fakirs who told me other rationales. One said the reason he created fakes was because they were the only thing he had ever been praised for — so realistic were his fakes that some considered him the Leonardo da Vinci of fake celebrity porn. He didn't do it for his own sexual gratification, or for the money. He did it simply for the comments and accolades others who frequented the forums gave him. He told me the popularity of his work made him feel not just accepted for the first time in his life, but respected. (And he wasn't alone – I've met other fakirs who told me similar stories.)

It seems odd, doesn't it? After all, the ethicality of the work is dubious – at best. But merely laying eyes on one of this fakir's creations would leave you in no doubt about the fact that he is an exceptionally skilled digital artist. A Photoshop master who creates altered images so lifelike they'd be indistinguishable from reality itself if not for his identifying watermark in the corner of the image giving it away as a manipulated photo.

The thing was, while his fakes of celebrities engaging in almost any sexual act you can imagine looked 100% authentic – like it was an actual record of the celeb caught in the act – a single image could take this fakir a full 24 to 36 hours of work in Photoshop. That's 3 to 5 hours of manual labour each day for a week straight – for one fake photograph.

I remember him telling me that the ultimate goal for many fakirs was creating fake movies of celebrities having sex – pasting their face over a porn star's face in a video. But think about the amount of work involved in a single Photoshopped photo: now imagine how much time would be required to create a short video clip, let alone a whole film. Indeed, while some fakirs did try to do just that, the results were always poor at best, manually created in programs like Adobe After Effects. They resulted in fake porn clips where you had what was essentially the digitally cropped, decapitated head of a celebrity – usually taken from the video of an interview on a late-night talk show – pasted directly over the head of the adult film star in a porn flick. The result was surreal and creepy: the celebrity's virtually floating head appeared to be happily chatting to some invisible person while the body the head floated on top of got roughly jostled by its sex partner.

'But fake celeb porn movies – authentic-looking ones – are never going to happen,' I remember the Leonardo da Vinci of fake celebrity porn saying. 'Unless you have a few million dollars and a team of skilled animators who can create a CGI replica of an actress's face and stitch that over the body of a porn star frame-by-frame, and then animate every facial expression to match what the base body is going through . . . well, even then you'd just be better off approaching that actress with the million bucks and paying her directly to make a real porn.'

And until 2017, he would have been right. You would have needed millions, skilled talent and lots of time to make a photorealistic fake celebrity porn video.

But then r/deepfakes happened, and everything changed.

In December 2017, I received a text message with a link from a friend of mine. His text read: 'Reminds me of your novel, but with videos instead of photos.' I clicked the link, and it took me to a *Vice* article[44] written by journalist Samantha Cole. Above the headline, the poster image showed *Wonder Woman* star Gal Gadot reclining back in bed with a big smile on her face. She lay there in short shorts, her bare legs exposed down to her toes, and a deep pink tank top, knotted up exposing her belly button. In one hand, she held a smartphone and in the other a black Fleshlight – a male sex toy. The image was a screenshot from a sex video ostensibly featuring Gadot. Below the image was *Vice*'s headline: 'AI-Assisted Fake Porn Is Here and We're All Fucked.' What Cole's article explained was that a person on Reddit was posting videos of female celebrities engaged in sexual acts – acts they'd never actually taken part in.

Sounds a lot like what some Photoshop fakirs had been trying to do over a decade earlier, right? The thing is these fake celebrity porn videos were leagues better – unimaginably so – than anything the old Photoshop fakirs could have dreamed up in After Effects. That's because they were not created by merely crudely cropping and pasting a celebrity's head from a source video onto a porn star's body in an adult video.

I'm sure you already know where I'm going with this: these videos were created using artificial intelligence. The digital face transfer was performed using complex algorithms to define the celebrity's facial structures as mathematical points which were then recreated and mapped onto the porn star's face in the adult video, giving the porn star the look of the celebrity while at the same time giving the celebrity's mapped face the mannerisms and expressions of those of the porn actress.

If this reminds you of the face swap apps from earlier, it's because these AI-generated fake celebrity porn videos *Vice* reported on were created using GAN-based technology – the same GAN technology that ultimately led to all those fun face swap apps.

Yet, these AI-generated porn videos were much too long and too advanced to have been created using any smartphone face swap app in existence in late 2017. So how was the Redditor making them? As *Vice* reported, the Redditor was a computer programmer who used open-source code from public machine learning libraries to create his (and it's pretty fair to assume that it is a 'he' simply because creating deepfakes celebrity porn is an almost exclusively male pursuit – almost) bespoke software that allowed its AI to learn the face of a celebrity and then replicate that face over the face of a porn star in an adult video. It's unknown just when this Redditor made his first AI-generated fake celebrity porn video but, by December 2017, he'd already created fake celebrity porn of Scarlett Johansson, Taylor Swift, Maisie Williams and Aubrey Plaza, in addition to Gadot.

So who was this Redditor? No one knows. Though the Redditor communicated with Cole, he declined to give his real name to avoid public scrutiny. Instead, *Vice* referred to the Redditor by his Reddit username.

That name?

'Deepfakes'.

And it's a name that the world has now adopted to refer to any video, photo or audio that has been altered or created using artificial intelligence. Before the Redditor chose his username 'Deepfakes' and founded the r/deepfakes Reddit community, AI-generated altered media was known as 'synthetic media' in research circles; in public circles – thanks to the popularity of all those fun smartphone apps – it was simply known as 'face swaps'.

Yet a unified label for AI-generated altered media was not the only contribution Redditor Deepfakes made to the field. His creations vastly raised the attention synthetic media received in the press, too – and it's the kind of synthetic media he made that sent the alarm bells ringing. Remember, before Deepfakes burst onto Reddit with his fake porn videos, the general public mostly saw 'face swap' tech as harmless and fun. Until then, AI-generated media had mainly been used for creating short, funny video clips with friends. But as soon as the technology turned towards porn, the panic started.

Before I go into why, it's important to note that it's not un-expected that synthetic media technology would be applied to porn in some way. After all, virtually every technological advancement has been applied to pornography. The printing press led to collections of erotic verse in the 1700s. Chemical-based photography led to Victorian still-image pornography in the 1800s. Celluloid cinematography led to adult films and adult movie theatres in major cities worldwide by the mid-1900s, and the advent of the VCR in the 1970s and the proceeding victory of VHS over Betamax allowed filmed pornography to move from those adult theatres to personal TVs in the privacy of one's own home by the 1980s. Why is VHS's victory over Betamax significant here, specifically? Because Betamax inventor Sony wouldn't license the technology to porn companies to distribute their adult content on, but porn companies were allowed to distribute their films on VHS. While cost was also a factor in VHS's victory, it has long been speculated the lack of Betamax pornographic content doomed the technologically superior format.

However, the most significant technological leap forward in porn's history came in the 1990s with the advent of the internet. Some figures suggest that in the early internet days – the mid-to-late 1990s – as much as 30% of all online content was pornographic, though hard data from this early time is difficult to come by. Regardless, the explosion of porn tube sites in the 2010s made porn easier to access than ever before. Of course, as more and more of our society, services, and media have moved online, porn's total share of web traffic has dropped, though it's still a significant factor in online activity.

According to January 2021 data[45] from web analytics firm SimilarWeb, two of the ten most-trafficked websites worldwide are adult websites: XVIDEOS took the #7 spot, and XNXX took the #9 spot. These sites are more popular than Amazon, Yahoo, Microsoft's Live and Netflix. Looking further, Similar-Web's data also revealed that four of the top twenty websites are adult sites, with Pornhub taking the #13 spot and xHamster taking the #18 spot. Each of these four adult sites receives more worldwide traffic than the likes of LinkedIn, Pinterest, eBay, TikTok, Microsoft and PayPal.

In 2019, Pornhub – just the third most-visited porn site on the planet, according to SimilarWeb's data, revealed that there were 42 billion visits to its website alone that year – 115 million visits each day.[46] And in case you think it's only men visiting these porn sites – think again. Figures obtained from 2014 data found that while 87% of male US adults aged 18–35 watched porn at least weekly, 28.5% of female US adults aged 18–35 did so as well – and virtually all did so online.[47] As a matter of fact, a 2017 study[48] of a decade's worth of data from Pornhub revealed that women spend, on average, one minute and fourteen seconds longer per visit browsing the tube site than men do. Yet both genders overwhelmingly love using the same type of device to consume their online porn. Eighty per cent do so via their smartphone.[49]

Pornography loves technological advancements because technology always propels the creation, distribution, and consumption of porn forwards. But if technology and porn have always been so inextricably linked, why did deepfake pornography generate so much panic as opposed to, say, the online streaming pornography found on tube sites, which is more pervasive and which, data shows, the average young adult is quite a fan of already? Or, in other words, why did *Vice*'s headline read 'We're All Fucked' when it came to AI-assisted fake pornography?

The issue comes down to consent.

While some may disagree with online pornography due to moral or ethical concerns, most publicly accessible online pornography consists of photos and videos of sexual acts between consenting adults. That consent is absent in virtually all deepfaked pornography.

In a deepfaked celebrity porn video, for example, there are always at least two people who may not have given their consent to appear. The first is the celebrity themselves. For example, Gal Gadot did not give her consent to appear in Redditor Deepfakes' AI-generated porn video. The second is the person originally in the destination video. The porn star, in this case Pepper XO, who starred in the destination video Redditor Deepfakes used to create the Gadot fake, did not give her consent for her body to be used with Gadot's face.

Redditor Deepfakes told *Vice*'s Cole he was just a programmer with an interest in machine learning – and that's what led him to create his custom deepfake software that allowed for the creation of the Gadot fake celebrity porn video. But when Cole asked about the ethical implications of how he applied the existing GAN technology he built his bespoke software from, Deepfakes dodged. 'Every technology can be used with bad motivations, and it's impossible to stop that,' he said and noted his software wasn't much different than the kind used by Universal Pictures to resurrect the late Paul Walker for scenes in the 2015 film, *Furious 7*. 'The main difference is how easy [it is] to do that by everyone,' Deepfakes said. 'I don't think it's a bad thing for more average people [to] engage in machine learning research.'

The Reddit website did think it was a bad thing, however. At least in the way user Deepfakes was applying the technology. In February 2018, Reddit banned the r/deepfakes community where user Deepfakes and dozens of others had posted their AI-generated fake celebrity porn videos. The reason for the ban? Reddit said it violated their existing 'involuntary pornography' policies. Other social media companies soon followed suit.

Yet by February 2018, when those bans began, Pandora's Box had already been opened. Seeing what Redditor Deepfakes did with his custom deepfake software based on existing open-source machine learning libraries, other programmers stepped forward to create their own deepfake software, too – but with one big difference. Their deepfake software allowed people with very little knowledge of programming or machine learning to easily create their own fake videos.

Weeks before Reddit shut down the r/deepfakes community, another Reddit user named 'deepfakeapp' had already created a new desktop-based software called FakeApp. FakeApp was built upon Redditor Deepfakes' bespoke deepfake algorithms. Yet, FakeApp packaged those complex algorithms into a user-friendly interface that people with little knowledge of programming or machine learning could use.

'I think the current version of the app is a good start, but I hope to streamline it even more in the coming days and weeks,' Redditor Deepfakeapp told *Vice* at the time.[50] 'Eventually, I want to improve it to the point where prospective users can simply select a video on their computer, download a neural network correlated to a certain face from a publicly available library, and swap the video with a different face with the press of one button.'

FakeApp is the program that led to the explosion of fancasting deepfakes on YouTube, including the Princess Leia one made by YouTuber Derpfakes I mentioned earlier. Before FakeApp, most YouTube fancast creators used their own custom code to make their deepfakes, too. It's just FakeApp wasn't only being used to create fun fancast deepfakes. Because of the app's relative ease of use, fake celebrity porn deepfakes exploded across the web in January 2018 – on Reddit first (before the r/deepfakes forum was banned) and then in internet forums that arose specifically dedicated to deepfake celebrity pornography. Unlike with the Photoshop fakirs of old, this time these forums weren't populated by skilled digital artists creating manually labour-intensive works, but rather by many who had zero previous machine learning or graphics skills and who barely needed to lift a finger to create something much more complex than a faked photo.

By that spring, the speed at which both fun fancast and fake celebrity porn deepfakes appeared was only matched by the iterations to existing deepfake software – as well as newer, even more powerful and user-friendly, desktop deepfaking programs that debuted later in the year.

Programs like DeepFaceLab.

The software launched in the latter half of 2018 and is available for free on GitHub,[51] a repository owned by Microsoft that houses open-source code the public can contribute to and use themselves. DeepFaceLab is currently billed as 'the leading software for creating deepfakes' – something which isn't just a blind boast. Thanks to its relative ease of use for even non-technical people, and consistent development and improvements, Deep-FaceLab is the go-to application for deepfake creators – both those you find on YouTube making fancast content and those who use the program to create fake celebrity pornography.

Considering it's, at best, charitable to say those creating fake celebrity porn operate in a morally and legally grey area, it should come as no surprise that it's hard to get a celebrity porn deepfaker to agree to an interview with a journalist. This difficulty was something that I experienced years ago, too, when trying to interview Photoshop fake celebrity porn creators for research for my novel – well, to a degree.

When I was writing *Epiphany Jones*, Photoshopped fake celebrity porn was still an outlier. That is, besides the people engaged in the actual internet subculture of celeb fakes, you rarely heard about it in the mainstream media. Perhaps that's why I eventually found several Photoshop fakirs willing to talk – back in the day there was no critical eye writ large on what they were doing. Or perhaps some people are more willing to talk to a novelist writing a fictional story about a man addicted to fake celebrity porn because they see novelists as more empathetic and understanding creatures than journalists. Or maybe it just came down to pride – after all, most of the Photoshop fakirs I tracked down had no hesitation boasting about the skill involved in creating their fakes. And while I can't condone what they were doing, as someone formerly enmeshed in the world of media, I can admit their skill was undeniable. Their sense that they were creating something with hard graft and technique wasn't misplaced, whereas with deepfakers now, well it's much harder to give them the benefit of the doubt.

So I can't blame celebrity porn deepfakers for not giving a journalist the benefit of the doubt either. Particularly after all those alarmist headlines. I spent months trying to connect with them. I joined online deepfake communities and sent private messages to members. I joined deepfake Discord groups and sent direct messages there, and I directly reached out to the few porn deepfakers I'd found email addresses for.

The first three months, I received no replies to my queries. None. At all.

Then, four months in, I got my first reply.

'Fuck off.'

That was in a direct message on a Discord server.

73

After getting a few more variations of 'fuck off' over the next few weeks, I received my first polite reply: 'Thanks, bro. But I'm good. Good luck with that.' That was followed by a few false starts with other porn deepfakers who ultimately weren't willing to talk about the porn part.

Finally, I got another reply. I'd recently sent tens of private DMs to deepfakers on one of the most popular online deep-fake porn communities. Some of these deepfake communities take the form of a tube porn site where deepfaked celebrity porn videos can be watched freely by anyone with an internet connection and without joining the site itself – just like you watch a video on YouTube. Attached to some of these tube deepfake porn sites are a public message board where, without creating an account, anyone can browse the dozens of forums and read what the sites' members talk about. If you create a free message board account on one of these sites, you're given instant access to post on the forums and send other forum users private direct messages. This is how I 'met' Pipohix, which, of course, is not his real name.

I originally contacted him because I found one of his latest fake celebrity porn deepfakes on the tube portion of the deep-fake site he frequents. The tube portion – just like YouTube does with standard videos – lets you search for deepfake porn by categories like Latest, Top Rated, Most Viewed and, unique to porn tube sites, Models (the category for the celebrities featured in, or the porn stars whose bodies are used in, the deep-faked videos). I clicked on 'Most Viewed' and was taken to a page featuring four columns of six rows of large thumbnails.

The #3 most-popular celeb porn video on the site that week was a deepfake titled 'Kate Upton Painal Casting 60fps'. It had under 100 likes at the time, yet over 21,000 views – and it had only been uploaded just days before. The video was over 22 minutes long, not unusual for celebrity porn deepfakes, but that's 100 times longer than your average Reface face swap video. This celebrity porn deepfake was also much more graphic than your average Reface face swap, even the ones sent to Faraz Ansari to bully them.

As the title suggests, the deepfake featured supermodel Kate Upton in an anal sex scene. Other fake celebrity porn deepfake titles all look fairly similar. Sometimes there will be a disclaimer of some sort; 'Not Angelina Jolie'; 'Fake Emily Blunt', but these are more to protect the deepfakers legally – or so some argue – than for the benefit of the viewer.

As for Pipohix's 'Kate Upton Painal Casting 60fps', when I clicked on the thumbnail, I was taken to the video's main page, where it was displayed in a large player window (again, if you know YouTube, you know the drill). I clicked on it because even from the small thumbnail, I recognised Upton's face. And just from that small image, it did look like Upton herself was bent over a desk, as a man behind her pulled her head back by her long blonde hair, so that she's forced to look directly into the camera. Her mouth hung agape, perhaps from pleasure. Perhaps.

Clicking on the play button, the deepfake commenced. It's 22 minutes of Kate Upton having sex. When the male adult actor in the video asked, 'Know what's about to happen?' Upton smiled at him, 'Mmhmm. We're about to get it on.' It's not her voice, her moans or her body, but it is her face. Well, her AI-generated photorealistic face masked over the face of the original adult film star whose body appears in the deepfake.

The description listed under the video announced that this time the deepfake wasn't a paid request, rather Pipohix made it for himself and the community. He wrote he's happy with the way the deepfake came out and he hopes viewers like it, too.

Besides Upton's face, everything else in the video is authentic. The destination video – the clip the celebrity's face was inserted into – is a fairly typical adult video you can find on any tube site. I don't know which porn production company produced the original because no original identifiers remain. Instead, Pipohix has added his custom watermark, along with a custom title card at the beginning of the video, signifying it's a deepfake and he is the creator behind it. One other watermark is in the video, too – the one the tube site places on all deepfakes uploaded to it, so viewers know where it originated from should the video be pirated by a competing site.

If you can somehow manage to look past the morally questionable ethics behind it, past the fact that nobody in this deepfake consented to its creation, it cannot be denied that, technically speaking, the video is very well made. It looks 100% real – most of the time. Occasionally in deepfakes like these, there are a few shots that, due to their angle, the face 'slips' just a couple of times. If you were watching it frame by frame, you could see it. But for the most part, if you didn't know deepfake technology existed (and many people still don't), you could be forgiven for thinking this was a real sex video starring Kate Upton.

Clicking on Pipohix's name below the video took me to his profile page, which lists all the celebrity porn deepfakes he's uploaded to the site. Only twenty at the time I found the Upton deepfake (some deepfakers have posted hundreds across various online forums), but they all looked done to the same technical proficiency. Four of Pipohix's twenty deepfakes were of Upton, five of the twenty were of actress Sofia Vergara and eleven of the twenty were of pop star Katy Perry.

Interestingly, Pipohix's profile page didn't only showcase his celebrity porn deepfakes. There were also over thirty photo albums, each dedicated to a different celebrity. Each album contained fake nude images of the celebrity. Here the photo albums were dedicated to a broader range of stars, including Tiffani Amber Thiessen, Selena Gomez, Denise Richards, Taylor Swift, Scarlett Johansson, Rihanna, Kristen Bell, Megan Fox and more. Some albums had up to twenty fake nude photos of the celebrity in question. This made me think that Pipohix was a Photoshop celeb fakir of old who transitioned to placing celebrities into fake porn videos when deepfake technology arrived.

I sent Pipohix a private DM via the site's forums. I told him I'm writing a book about deepfakes and I hoped he'd agree to chat because he is clearly skilled and technically proficient at deepfaking – remember, I'd already been told to fuck off several times. I gave him an email address I set up to use to register on the forum, should he want to contact me that way. Based on my months of dead-ends and run-arounds with porn deepfakers, I didn't hold out much hope for a response.

Four days later, I received one.

His reply was brief but surprisingly friendly compared to communications with other deepfakers I'd had. He thanked me for the compliment and said he's interested in doing the interview provided he remains anonymous. He said he'd prefer to do the interview by email and gave me his.

I sent him an email to the address specified, along with my questions, assuring him I will respect his anonymity. I would make an alias for him, such as 'A deepfake creator I'll call "David" in order to protect their identity.'

Later that day, he replied, telling me it's OK to refer to him by the name he goes by in the deepfake community – Pipohix.

He said he was just originally frightened that I would want to know or use his real name. His candidness here again surprised me. Here was someone who was not defensive in the least – a nice change from some other interactions I'd had with deep-fakers. He ended this message saying he'll try to answer all the questions I sent him within a few days, and he signed off by wishing me a nice day and that I stay safe – a sadly all too common parting wish in the pandemic era we were living in.

Turns out I wouldn't need to wait a few days to hear from him again. The next day he came back with complete answers to all my questions. Following them was a short note apologising in advance if I didn't understand the way something was written as English is not his first language.

I realised from his response to my first question that I was right: Pipohix is a Photoshop celeb 'fakir' of old. Yet he got his start in photo manipulations, not because of women, but because of cars.

Pipohix wrote that in the early 1990s, thanks to saving for some time, he bought his second car. This was a big accomplishment for him, so he immortalised his purchase in some photos. However, when going through them he noticed each photo had objects in it that detracted from the car itself – traffic signs, pedestrians, or just a non-visually appealing backdrop. Wouldn't it be great, he thought, if he could edit those unwanted objects out of the photos of his new car?

Enter an old Windows program called PhotoFinish. It was considered a pretty advanced photo editing application back in the day, though by today's professional photo editing application standards, it might as well be from the Stone Age. Still, it did the trick for Pipohix. After using PhotoFinish to clean up the colours and edit the background, he then copied his car out of a photo altogether and dropped it into other photos – photos that provided a better backdrop for his recent acquisition, like one featuring a lovely villa.

Having mastered the art of placing automobiles into photos they never originally appeared in, he moved on to adding other elements into the photos, such as a woman in lingerie reclining on the hood of the car. Pipohix says he considered this his first adult photo creation.

Shortly afterward, the internet began to become commonplace in the country where Pipohix is from. One day while browsing the then-new worldwide web, he came across a faked photo of Victoria Principal, the American actress best known for her role as Pamela Barnes Ewing in the classic television series *Dallas*. Since he already understood the basics of early digital photo manipulation, he attempted to create his own faked photo – though the initial results weren't the best, he admits.

Undeterred by his unsatisfactory initial attempts, Pipohix said that after a while, he started to become more skilled at image manipulation. Eventually, he shared one of his creations to a newsgroup – the early precursors to today's online message boards. Surprised by the amount of reactions to his first post he created another fake. That's when Pipohix went in search of burgeoning online forums and found other photo fakers, trading tips and tricks about the craft with them. Eventually, Pipohix traded the PhotoFinish app for the industry-leading Adobe Photoshop and with this more advanced tool, his photo fakes – both those he kept private and those he shared online – became more professional looking.

Fast forward to 2018. Pipohix was on YouTube one day when he came across a fancast deepfake that featured actor Nicolas Cage's face inserted onto a woman singing the Gloria Gaynor classic 'I Will Survive'.[52] Before then, Pipohix had heard some things about deepfakes and the ability of artificial intelligence to alter video footage, but the real implications of the tech never occurred to him until he saw Cage, as a woman, belting out the classic a cappella.

'I watched it a number of times in a row and suddenly understood the possibilities,' Pipohix wrote. '[Deepfake technology] was so new and uninhibited that this was possible for ordinary home computers. It's like having a Hollywood studio on your desk.' But a Hollywood studio in your home would still cost millions, and, as Pipohix points out, all you need to make deepfakes is a normal PC. 'The only limit is your imagination and of course your skills, but you can do something about that.'

Speaking of which, it's important to quickly note just how desktop deepfake software, like DeepFaceLab, which deepfakers use to create both fancast and porn deepfakes, is different than the face swap apps we use on our phones.

Deepfakes of the quality and length that fancast and fake celebrity porn videos usually are would be impossible to pull off using a smartphone app. The reason why comes down to computing power. Because of both the machine learning and video rendering processes involved in making a lengthy deepfake, a significant amount of processing power is needed – with a heavy emphasis on graphics processing power. Today's smartphones just don't have that power. Thus, creating the deepfake that entertains, titillates, or worries everyone currently requires a PC with a powerful CPU and an even more powerful GPU, or dedicated graphics card (that's the piece of hardware that makes video games run smooth and crystal-clear on desktop computers).

Also, unlike touch-friendly face swap smartphone apps, DeepFaceLab doesn't have much of a graphical user interface. It's little more than a collection of folders with batch files inside. Upon launching the program, a user will select the 'source video', which has the face of the celebrity they want to extract, and the 'destination video', which contains the scene, and the body the celebrity's face is to be mapped onto.

If you are creating that deepfake of Kate Upton, for example, your source video would be a clip of Upton – perhaps from a catwalk or, even better, one of her giving an interview. Interview clips generally show the celebrity's face going through a wider range of expressions from many angles, which helps the AI learn what the celebrity's face should look like all around. The destination video, in this case, would be the original adult video with the porn actress (the 'donor body' as it's known in deepfaking circles) bent over the desk.

After defining the source and destination videos in Deep-FaceLab, the user launches another batch file that will create what is known as a 'dataset' or 'faceset' of the celebrity to be deepfaked. DeepFaceLab creates this faceset by extracting all the frames the celebrity appears in from the source video and saves them as individual image files. Remember, every second of video is made up of around 30 frames – so that's 30 still images for every second of source video of the celebrity you have. If your source video is 10 minutes long, that will give you roughly 18,000 images of the celebrity's face. The DeepFaceLab user performs these faceset operations on the face belonging to the donor body in the destination video as well. And the more images of the celebrity's and donor body's face you can add to the faceset, the better (you'll soon see why).

But first, the next step in DeepFaceLab is to manually out-line the celebrity's and donor's faces in a few of those still im-ages that make up their respective facesets. This outline tells DeepFaceLab what portion of the faces you want to replicate and replace. You can include anything from just the front por-tion of the face (forehead to chin) to the whole head (includ-ing ears and hair).

Once both faces are outlined, the user launches another batch file, which tells DeepFaceLab to analyse the thousands of still images in both facesets – concentrating on the areas outlined on the respective faces. The more images you have, the more analysis the software can perform. The software uses its analysis of the faces to build what is called a 'model'. This model is what the AI will use to generate its own artificial version of the celebrity's face and accurately map it onto the face of the donor body – while preserving the original expressions of the donor's face (such as showing that face moaning in ecstasy). This portion of the process is known as 'training'. That is, the AI is learning from its examination of the facesets to realistically reproduce the celebrity's face combined with the donor's facial expressions.

Training models can go through hundreds of thousands of iterations. This is the back-and-forth GAN game of forger and inspector we talked about earlier. So in our Upton example, while in training, for every iteration, one AI tries to forge a perfect copy of Upton's face based on the faceset extracted from the source video. At the same time, the other AI inspects that forgery and marks it as genuine or fake. The more itera-tions, the better the forger AI gets at knowing where it's failing to fool the inspector and makes alterations to its next attempt accordingly. In other words, with every iteration a replica of the celebrity's face is produced that is only more and more lifelike.

Even for a PC with a reasonably powerful CPU and GPU, these iterations can take days. For example, on Day 1, a PC with a moderately powerful graphics card might generate 50,000 iterations (attempted individual forgeries of the celebrity's face) during the ongoing training of the model. By Day 2, 150,000 iterations may have been generated. By Day 3, you could have 300,000 iterations. One of the ways to cut down on this lengthy training time is to have a computer with a more powerful graphics processor.

A DeepFaceLab user can pause the model's training at any time to see the most recent iteration of the celebrity's face mapped onto the donor body. If it looks real enough to the deepfaker, he can stop the training right there, make a few manual adjustments to the model if desired and then merge the forged celebrity face onto the donor body. All the deepfaker needs to do then is export the porn video with the merged celebrity's face on the donor's body and, hey presto, you have a deepfake.

In 2018, Pipohix's PC just happened to have a CPU and GPU with enough processing power to run DeepFaceLab – well, barely. So, after seeing the female Nicolas Cage singing 'I Will Survive', Pipohix downloaded DeepFaceLab and gave it a go. Yet, at the time, DeepFaceLab was still a new, unrefined application that was much more difficult to use than it is today – especially for someone who, while they may be a skilled and proficient digital artist, was not a programmer.

'It was too much and I could not oversee so much data and JPGs,' Pipohix confessed. 'Also smoke was coming from my computer because [DeepFaceLab] requires so much computing power from the PC/GPU. I ended up tucking it away with the idea of checking it out later.'

Yet later may have never come had it not been for the global Covid-19 pandemic that consumed the world in 2020. With lockdowns in his country, and the resulting free time, Pipohix decided to give DeepFaceLab another try. By 2020 the program had advanced considerably and there were also much better online documentation and tutorials that now existed that made it easier to learn and master.

When he replied to my questions, Pipohix had been making deepfakes for less than a year. Still, his were some of the best deepfakes I'd seen yet – and by 'best,' of course, I mean some of the most realistic-looking to my eyes; the kind it's virtually impossible to tell it's a fake video.

I asked Pipohix how he chose which celebrities to deepfake.

'Everyone starts with a celeb you like,' he said. 'In my case Katy Perry and Sofia Vergara.'

He concentrated on making deepfakes of those two until people began contacting him for customs. That's 'custom' deepfake work – work that non-deepfakers are willing to pay someone proficient in DeepFaceLab or FakeApp for to create their favourite celebrity starring in the porn of their dreams.

But when discussing custom deepfakes made for other people, Pipohix is very pragmatic. He says that he at first refused the work. Why? Deepfakes take a long time to make – days and days for a good one. If you spend days dedicating your computer's load to creating fake videos of actresses you aren't into, what's the point? To Pipohix deepfaking is a hobby. Fake celebrity pornographers get into the craft to fake *their* famous crushes, after all.

But then Pipohix got an email from someone requesting a deepfake of Kate Upton. And that's something that piqued Pipohix's interest because Upton was one of the celebrities he planned on deepfaking anyway. He took the job. Since then he's made several other custom deepfakes, with prices starting at around $60 USD, though some of his custom deepfake work can go for double that. It all depends on what the request entails — if it's possible and how much work it will take.

That possibility relies on exactly what the customer wants the celebrity to look like and to be doing in the fake porn video. For example, the customer could want something particular: like the celebrity's face mapped onto a specific porn star's body, in a specific porn video. Other times, customers will just say they want a particular celebrity in a, for example, masturbation scene. It's then up to Pipohix to find source and destination videos that fit.

Given his 'Kate Upton Painal Casting 60fps' deepfake hit the #3 spot on the tube site's charts just days after he uploaded it — and he continues to get requests for custom deepfakes — many clearly appreciate his work. I asked him why he thinks people are attracted not just to deepfakes, but his deepfakes, specifically.

Again, Pipohix's reply was surprisingly candid. 'I have no idea,' he admitted. 'I make a deepfake that I think is fun and try to deliver some quality. Luckily a lot of people like what I like too . . . Moreover, I try to deliver something original, as far as possible. [I] don't think we're waiting for the 100th Emma Watson standard masturbation fake or missionary position.'

This last comment is, I have come to realise, something of an in-joke among the celebrity porn deepfake community. By most accounts — and by the tube site's rankings — *Harry Potter* star Emma Watson is the most deepfaked celebrity in the world. The specific tube site Pipohix uploads his deepfakes to hosts more deepfakes of Watson than any other celebrity. There's even a year's-old forum thread on the site in which the original poster laments that Watson has 50 deepfakes already while many other celebrities have just one. At the time of the original post, there may have been 50 deepfakes of Watson, but at the time of this writing there are over 340. The next-closest is Scarlett Johansson at just over 280.

Watson and Johansson are among some of the most Photoshopped fake nude celebrities, too, as I discovered in my research for *Epiphany Jones* all those years ago. As a matter of fact, many Photoshop fakirs could hardly wait for the day Watson turned 18 – which brings me to another important observation in my research into fake celebrity pornography. Despite the obvious ethical boundaries fake celebrity porn creators traverse, any fake celebrity porn community I've ever come across has always had the same two strict, non-negotiable rules:

1) No fakes of any celebrity under the age of 18.
2) No fakes of any celebrity over the age of 18 using their face or body from photos or videos taken when they or the donor body were under the age of 18.

Now, this is not to say fakes of underage celebrities have never been made, but the website Pipohix uploads his deepfakes to, and all others I've found, adhere to these rules – and crossing them will get you immediately banned and ostracised from the faking community. Not only that, all fake celebrity porn communities I've found adhere to standard adult website practices, allowing any visitor to easily flag and report photos or videos that might feature someone underage. This goes for sites hosting deepfaked celebrity porn videos as well as for the old Photoshop fake celebrity porn forums.

Speaking of Photoshopped celebrity porn, as he's a fakir of old, I wanted to get Pipohix's opinion on something. As I've already mentioned, no matter what one thinks of the practice, it can't be denied that creating fake nude celebrity images in Photoshop, like creating anything else in the program, takes artistic skill – the blending of colour and light and shadows using digital tools controlled by the human hand and overseen by the human eye. Photoshop may be the tool, in other words, but the quality of the output is down to the human's talent. But the power and draw of deepfake technology are that human skill isn't really required. The most manual, human-labour intensive task in DeepFaceLab deepfake creation is the outlining of the celebrity's face to tell the AI what part of the source video to study to learn how to forge it.

So does Pipohix feel like he's doing a fair amount of the work when he makes a deepfake, or does DeepFaceLab's AI get all the credit? If not, what share of the credit does he feel he deserves for his popular deepfakes?

'I would say slightly more than half,' Pipohix answered. 'I determine which celeb will be in my chosen video and the quality of it. The program only executes the computing power. But my half is decisive, otherwise all deepfakes would be the same.'

While I would give DeepFaceLab's AI a more significant share of the credit for the final fake porn video's realism, I do take Pipohix's point. DeepFaceLab, and deepfake technology in general, are tools without objectives at the end of the day. They may be advanced tools based on artificial intelligence, but no deepfake AI I've heard of has ever started creating fake celebrity porn videos of Kate Upton without a human telling it to do so. Deepfake technology does not have its own intent or desires; it is the people who are using it who do.

Pipohix also supports his rating by noting his old school skills as a digital artist do still contribute much – at least to his particular deepfake workflow. While DeepFaceLab's AI does do the actual generation of the face, deepfakers can manually tweak the replicated celebrity face during the merge stage – such as changing its size or sharpening or blurring its edges – before it's set in place on the donor body. Though this step isn't necessary, I can see how digital artists used to manually tweaking elements on their screen may play around with such tools more than deepfakers with no digital arts background.

'In retrospect, my experience with Photoshop comes in handy,' Pipohix noted. 'I quickly see with a graphical eye what is wrong [with a deepfake] and what needs to be done [to fix it].'

Still, artificial intelligence that places celebrities into ultra-realistic-looking pornos is *leagues* beyond the manual editing tools of yesteryear that allowed for the creation of fake celeb photos – no matter your experience with those digitally archaic tools. The fact that AI is even being used to create porn of famous people at all . . . every time that really sinks in I can't help but shake my head as I think back to what the 'Leonardo da Vinci' of Photoshop fakirs told me all those years ago: 'But fake celeb porn movies – authentic-looking ones – are never going to happen.'

Never is a lot shorter than it used to be.

Is deepfake technology a dream come true for fakirs? That's what I asked Pipohix. After all, the ability to create fake celeb porn videos had often been referred to as the ultimate goal; the holy grail.

'If there was no deepfake technology, there would be no fake celebrity porn videos,' he replied.

Well, maybe no realistic-looking ones.

The reason?

Mathematics.

Pipohix pointed out that even a short video (porn or otherwise) could be made up of 30,000 frames – each frame being a still image. In this case, a 16-minute porn video, which is shorter than Pipohix's 22-minute Upton opus, is made up of around 30,000 images/frames. Before deepfake technology, if a Photoshop fakir wanted to create a *realistic* 16-minute fake celebrity porn video, they would have needed to manually edit all 30,000 individual frames, one at a time – each taking hours, maybe days, in Photoshop. In other words, it would be a years'-long process to make a lengthy realistic fake celebrity porn movie by manually editing each frame, and as Pipohix notes, the celebrity in the video would likely be forgotten by the masses before the deepfake was even finished.

He's not wrong. After all, the saying 'fame is fleeting' exists for a reason. Research conducted in 2020 revealed[53] that 70.1% of all actors who received one film credit in a movie between the years 1949-2019 never received a second credit. 28.2% of actors received between 2 and 20 credits. But only 1.8% of all actors ever receive more than 21 credits. In other words, the majority of Hollywood actors will appear in a few films and that's it. The Hollywood machine seems happy to discard its up-and-comers the second it finds someone more bankable. To see how true this is, you simply need to browse Netflix for movies or TV shows that are older than five years. You'll soon hear yourself saying, 'I completely forgot about that actor … what ever happened to them?'

As for whether deepfake technology is a dream come true, Pipohix admits it is for some, but for celebrities, he also admits, it's a nightmare.

And it's that admission I now had questions about. I responded with two follow-up questions. The first was how would he feel if he found a deepfake of himself? The second was if there was a limit to the types of pornographic deepfakes he would make – for example, would he make deepfakes of non-celebrities?

Pipohix answered that he only fakes celebrities and that he wouldn't mind if a deepfake of himself spread around the web.

But is that the case for all these celebrity pornographic deepfakers? Would they all really not mind if it happened to them? It's a question I wanted to put to others. But that wouldn't be easy.

On the same day I originally sent Pipohix my follow-up questions, another deepfaker I'd been working to interview sent me a new DM on the community both they and Pipohix frequent. This deepfaker, who had previously agreed to an interview as well, now wrote to ask which other deepfakers I reached out to on the forum we're communicating through.

I replied that I contacted quite a few, knowing few would respond, and that I wouldn't reveal the names of the ones who had.

After I sent my response, I didn't hear back from that deepfaker again.

In the time since, I've also put my question of how would a deepfaker feel if they found a deepfake of themselves to a few deepfaking communities on different platforms. In one I received no replies. In another I was booted from the community.

That wasn't exactly unexpected. In the past, I've interviewed other types of people who operate in, well, questionable areas. And when you start asking questions relating to the ethicality of what they do, well is it any wonder some of the deepfakers I reached out to simply replied, 'Fuck off'?

But several weeks later I did receive a message from another celebrity porn deepfaker who was a little more forthcoming about the fact that, yes, what these types of deepfakers are doing isn't exactly harmless. This was one of the deepfakers I had reached out to earlier, asking if they'd be willing to do an interview. The unexpected thing about this deepfaker, though, which I didn't know until they replied, is that they're … a woman. A *female* celebrity porn deepfaker.

As I noted earlier, data shows over a quarter of women admit to watching porn on a regular basis – and I've interviewed female porn stars and sex workers before who can eloquently express why they've chosen to work in the porn and sex industries. So I understand that porn is something that is not wholly male-driven or initiated. Yet in all my years of researching Photoshop celeb fakirs, I had never met a female one. Until I received this particular deepfaker's message, I'd assumed that pornographic celebrity deepfaking was *naturally* something only men who are obsessed with certain movie stars did.

I was wrong.

Her message to me began with a note saying she likes to keep a low profile, so won't do an interview, but she would like to make her own statement regarding her perspective on deepfaking. First, however, she asks me not to identify her by the name she uses on the deepfake community we're messaging each other through. Instead, for this book, she asks that I refer to her as 'SpicyDeepFaker'. It's definitely a name that matches the tone of her message.

SpicyDeepFaker is fairly prolific. She's posted dozens upon dozens of female celebrity pornographic deepfakes – way more than Pipohix. She also puts a lot of design into her deepfakes' thumbnails, complete with her personal logo, which would make a creative director hire her in an instant. Several of her deepfakes have over 100,000 views and one of her most popular has almost a quarter of a million views and a massive amount of likes, making it by far one of the most popular deepfakes on any site I've seen.

As for SpicyDeepFaker herself, she says she is a twenty-one-year-old woman of Latin American origin. She says the reason she likes to make deepfakes is to appreciate the technology that lies behind it, while admitting that she does take donations as her lifestyle is expensive. Her profile on the community she frequents notes she does do paid custom deepfakes, but her customers better have cryptocurrency because that's the only payment she'll take. But in her message to me, she's quick to note she has limits to who she'll deepfake and in what scenarios. She follows the rules of the deepfaking community. No revenge porn. No porn of people under the age of 18. No bestiality. That being said, she admits she likes experimenting when it comes to creating deepfakes of celebrities.

SpicyDeepFaker believes that the underlying technology behind deepfakes will continue to improve exponentially and the creators and viewers of deepfake pornography will only enjoy more anonymity in the future thanks to the increasing adoption of anonymous payment systems like Bitcoin. She also alludes to Rule 34, which to me suggests she is steeped in internet culture. If you haven't heard of this maxim, it's essentially a much less-wholesome version of Disney's 'If you can think it, you can do it!' Rule 34 basically states: if something exists, then somewhere on the internet you can find porn featuring it.[54] I'm sure you could spend an entertaining afternoon testing this theory out if you don't want to take my word for it, but spoiler: it's mostly true.

So to me it seems that for SpicyDeepFaker, AI-generated fake celebrity porn is a construct that was inevitable sooner or later. She writes that as long as people have perverted desires and fantasies about their celebrity crushes, deepfakes will only prosper.

And that's just the thing – celebrities, sexual fantasies and perverted desires – well, it's hard to see how any of those are ever going away. So, according to SpicyDeepFaker's logic (which seems pretty spot on), celebrity pornographic deepfakes aren't going anywhere.

As that's the case, what I want to ask her – all of them, really – is, what about the harm they do? What about the distress they cause to these celebrities, who have not consented to appear in these deepfakes? The very same deepfake celebrity porn that some deepfakers are profiting from? But SpicyDeep-Faker's message comes to a close musing that there is a price to being famous. She believes that price is privacy, and now in a world where deepfakes exist, the price also includes the trading of their core identity.

I know actors who would strongly agree with SpicyDeep-Faker's assessment – and most would probably admit that their core identity doesn't only get traded in the world of deepfakes, but in the world of Hollywood, too. Yet at least in Hollywood they've consented to that trade.

But still, I didn't have any additional answers to that other question: how would these celebrity porn deepfakers feel if they found someone had made a deepfake of *them* without their consent?

While most deepfakers may not be willing to offer an answer to that question when approached by a journalist, it is a topic that has been discussed on celebrity deepfake porn forums. On one such community, there's a thread started by a deepfaker who laments that his girlfriend found out that he creates deepfakes and she thinks it's wrong. She asked how he would feel if he found a deepfake of himself, so he, in turn, put that question to the greater community.

The replies, of which there are almost two dozen, are varied. One said they would feel neutral about it, while another said if they found a deepfake of themself or their wife and didn't like it, they would ask for it to be taken down. But another wrote that they wouldn't care at all if they were deepfaked; that they'd probably find it flattering and hilarious that someone wanted to see them in porn. Yet, they added that creating deepfake porn of normal people is out of bounds. Another insisted deepfake porn is 'art'. One wrote celebrities should take it in their stride, asserting celebrity deepfake porn is no worse than crazy fan-fiction people write. Another, still, was more blunt, stating they didn't feel sorry for celebrities who complain about their likeness being used inappropriately, arguing that the celebrity might as well be demanding that someone can't fantasise about them.

Keep in mind this is just a summary of the comments of a few deepfakers who had no problems speaking publicly about the ethical implications of celebrity porn deepfakes. Is this a representative sample of all celebrity porn deepfakers? It wouldn't be fair to anyone, the deepfakers or the celebrities they fake, to make that assumption. What I can say, with certainty, is the above examples came from fewer than 10 of that community's hundreds of thousands of members.

It should also be noted that the term 'celebrities' here is used colloquially to describe the types of people who are deepfaked into the videos posted on deepfake porn sites. However, 'celebrities' doesn't only mean movie stars and supermodels. The sites that host deepfakes of Emma Watson, Scarlet Johansson and Kate Upton, also host them of social media influencers, like fashion and make-up YouTubers, Twitch videogame streamers, and politicians. One politician, in particular, seems to be a very popular target for pornographic deepfakes: Alexandria Ocasio-Cortez. Better known as 'AOC,' she has served as the progressive Democratic US Representative for New York's 14th congressional district since 2019. She is beloved by the left and loathed by many on the far right.

If you thought that shallowfakes were a problem for politicians, at least they pale in quality, length, and explicitness compared to pornographic deepfakes. And while it's true the average politician is more likely to be the victim of a shallowfake, for AOC it's different: on one popular website there are around fifty pornographic deepfakes of Ocasio-Cortez alone – more than most Hollywood celebrities have.

And I wish I could say I was surprised, but it's not just well-known movie stars, influencers and politicians who are being deepfaked either. Far from it.

In the spring of 2019, a new website appeared. It was called DeepNude.com, which also happened to be the name of the software it hawked. DeepNude was a program available for Windows and Linux based on GAN-style deep learning libraries. However, unlike popular desktop deepfaking programs like DeepFaceLab, DeepNude applied GAN technology in another way. You see, while DeepFaceLab enables users to place a person's face onto another person's body in any video, DeepNude left the target's face alone. Instead, what DeepNude allowed for was the ability to digitally strip the clothes off the person in any photo the app processed.

Have a photo of Emilia Clarke in her Daenerys garb from *Game of Thrones*? Process it through DeepNude, and the clothes come right off, revealing what Clarke's nude body looks like underneath the outfit. Of course, it's not *actually* Clarke's breasts

95

or genitals, but AI-created, photorealistic approximations of them based on the fit of the outfit she wore.

In other words, with DeepNude, the face stayed real, while the nude body was fabricated.

But why was DeepNude, which only worked on still *images*, more alarming in some respects than DeepFaceLab, which allows for the creation of realistic fake porn *movies*?

First, it may have only worked on still images, but it was incredibly simple to use, meaning the app had virtually zero learning curve. There was no messing with defining source and destination videos, creating facesets or training models. Deep-Nude simply needed to know what photo to go to work on – all the users had to do was select it, and DeepNude did the rest in virtually no time at all. Simply put, anyone could now create fake porn images out of a single photo within seconds of downloading the program.

Second, while DeepFaceLab seems to be a fake celebrity porn deepfaker's program of choice, it isn't solely designed for creating fake porn. It's also the program used to create many of those entertaining YouTube fancasting videos. And while DeepFaceLab is definitely the go-to program for creating deepfake porn, it isn't explicitly designed to create fake porn of *female* celebrities only. While the overwhelming majority of celebrity porn deepfakes created with DeepFaceLab target women, in the last six months there have been an increase in the number of male celebrities featured, too. A quick browse of one popular deepfake website reveals that alongside Kate Upton and Gal Gadot, there are celebrity porn videos of pop star Nick Jonas, social media influencer PewDiePie and a host of Hollywood actors who play superheroes, including Hugh Jackman (Wolverine), Henry Cavill (Superman), Chris Evans (Captain America), Tom Holland

(Spider-Man) and Chris Pratt (Star Lord). In other words, Deep-FaceLab is an equal opportunity deepfaking software. Deep-Nude, on the other hand, was *specifically* built to create fake porn – and fake porn of a certain type of person only.

Should you think that DeepNude's AI simply stripped the clothes off of anyone in a photo that was processed through it – think again. The software *only* worked on pictures of females.[55] It did not remove the clothes from men whose photos were uploaded into the app. DeepNude, in other words, was the first deepfake software in history specifically built to be weaponised against women. And not just celebrity women. *Any* women. A work colleague, a stranger, a mother, a daughter, a wife. You. Any woman in the world could now be stripped naked by artificial intelligence in an instant. And the resulting fake nude image of that 'regular' woman could then be shared however the user liked, or just saved to their private collection.

DeepNude operated in semi-obscurity for several months until multiple publications reported the program's capabilities in late June 2019. Within twenty-four hours of the reports, DeepNude's creators shut the software down. After the media attention, the program's creators posted a message to Twitter explaining the cessation of operations.[56] 'We created this project for user's entertainment a few months ago . . . We never thought it would become viral and we would not be able to control the traffic.' The message went on to reveal that its creators were pulling the software: no one else could download it. 'The world is not yet ready for DeepNude,' it ended.

But ready or not, DeepNude was here to stay. Within days of the app being removed from the creators' depositories, copies of its code were spreading across the internet. Within weeks, DeepNude's code was used as the basis to create other apps – this time web-based – to which anyone could simply upload a woman's photo to have her clothes stripped clean off. This eliminated another barrier of entry to fake pornography: you no longer needed to download and use an app based on deep learning processes to create fake porn; you no longer needed to download any app at all; you only had to upload a photo to a website. And even that changed: soon enough, even a website wasn't necessary.

In October 2020, Sensity, the company of artificial intelligence and machine learning expert Giorgio Patrini, released a report[57] that seemed to justify all those alarmist deepfake headlines we've talked about. Sensity is one of the few companies in the world created specifically to monitor and help others mitigate the growing threat of deepfakes. Any stats you've seen mentioned in the media about deepfakes likely came from the data Sensity collects and releases in its reports.

So, what was so shocking about the report Sensity released in October 2020?

It revealed that all anyone wanting to deepfake a woman needed to do now was text a photo of the victim to a chatbot in the popular Telegram messaging app. The bot would then strip the woman of her clothes and send the faked nude back to the person who'd requested it.

Sensity's report revealed that the bot was an evolution of the DeepNude software that had been shut down the year before. Though the bot wasn't the creation of the people who made DeepNude, it built upon its legacy. Since Sensity's monitoring of the bot began, by July 2020 it had been used on over 104,000 women, all of whom were digitally stripped, their resulting images shared in an associated Telegram channel with tens of thousands of subscribers. That meant that subscribers could see the digitally stripped women other users had the bot create. In other words, over a hundred thousand women were having their artificially generated nude bodies, with their real faces, paraded in front of tens of thousands of strangers every day – all within a simple messaging app.

And those hundred thousand are just the victims that had their nude bodies *shared* with other users of the channel. Further data Sensity uncovered showed the bot had been used to strip at least 680,000 women in total. Additionally, the number of images processed by the bot grew almost 200% in the final three months Sensity's report covered – meaning it was only becoming more popular. Furthermore, the bot and its associated Telegram channels had over 101,000 users worldwide – and 70% of the photos the bot 'processed' (i.e. stripped of clothes) were of ordinary, non-celebrity women.

As easy as the bot was to use, it was also free to use. Anyone could submit ten photos of women to be stripped *per day*. And, if the user so chose, they could pay a fee between $8 and $12 to remove watermarks from the fake nude images the bot created for them. As bad as that was, it gets even worse. In monitoring the bot, Sensity believes they came across some deepfaked nudes that may have been generated from images of minors. It was with the discovery of these images that Sensity contacted law enforcement.

All this seemingly birthed from a simple deepfake program that was pulled from the internet a year before.

This Telegram bot is just the tip of the iceberg though. In researching this book, I've crawled many pornography image-boards on the web looking for deepfakers to contact. In recent months, I've increasingly found ever more uploads of deepfaked still images. These are deepfake nudes created by apps that are likely based on DeepNude's code. They are collected into albums the uploader has made. The albums have names like:

'Deepfakes of my friends'

'My busom history teacher deepfaked'

'Deepfakes of sexy girls I know on Instagram'

These are albums anyone with an internet connection can view.

And it's not just imageboards.

As I'm writing this, I've just logged into a deepfake Discord server where someone has announced they've launched a new web-based app that allows people to upload photos of the girls they know to see them stripped nude.

Another person on the server asks if anyone wants 'to buy a DeepNude software?' They say they'll sell it cheap, but they only take PayPal. But if you buy it, you'll be able to create an unlimited amount of high-quality 'deepnudes' so long as you have pictures of girls in bikinis, which the software will strip from them.

Several messages above this, another user chimed in earlier with the announcement that they are running a holiday sale. Until the end of the month over two dozen deepfaked photos only cost $12. The person offering the sale reminds interested parties that it's important to submit pictures of women in bikinis in order to get the best results.

But it's not just on online pornography forums and Discord servers dedicated to deepfakes that you see references to deepfake porn increasingly appearing. On the world's most popular online discussion site, I find a public post by a user that describes how he was bullied by some classmates in high school. They would sometimes grab him from behind and he would feel his bullies' erections pressing against him. Now a decade later, the poster says he stalks his former bullies on social media and uses photos they've posted to swap their faces onto images of naked men, which he then masturbates to. He ends the post by asking readers if they think that's weird.

Is that weird?

I think that's the wrong question to ask.

Everyone has their kinks – and I don't think anyone should ever be shamed for those kinks – but, generally speaking, legal sexual kinks, when represented in photographic form, have historically involved consenting adults. When you are using someone's face without their permission, as you are in a pornographic deepfake, there is no consent. So, perhaps the proper question to ask is not is it weird, but is it *right*?

It's a question some celebrity porn deepfakers do ask themselves. And most justify their continued deepfaking by saying their deepfaked porn content is clearly identified as being just that, fake. Deepfaked celebrity porn uploaded to popular deepfaking websites generally do this by watermarking the videos and/or placing the qualifiers 'not' or 'fake' in the title of the video. Even DeepNude and the Telegram bot watermarked their faked nude images. However, watermarks can easily be removed before the image is shared again. Even if the watermark is left on, that doesn't address the ethical issue: people may know a nude image or pornographic video is fake, yet it is still humiliating to the victim in that deepfake. And such humiliation can leave long-lasting scars. Just ask anyone who's been bullied.

Speaking of scars, it was my objective in exploring the pornographic implications of deepfake technology for this chapter to try to get the perspective of the deepfakers themselves who create the non-consensual porn. Is it always just for their own sexual gratification? I would never want to presume.

When I first started research for *Epiphany Jones* all those years ago, I hadn't yet decided whose eyes I would tell the story through. A trafficking victim's? A trafficker's? A Photoshop fakir's? I ultimately settled on an active consumer of Photoshopped fake celebrity porn after researching various kinds of sex addictions and why those addictions arise (almost always from trauma or abuse). In my novel, the main character began consuming fake celebrity porn in an attempt to numb his anguish over a sibling's death. It was an attempt to distract from his hurt. And in talking to sex addicts and old-school Photoshop fakirs, I've learned that you never know *why* people are engaged with the vices they are. I now understand that some have addictions to various types of pornography, whether they are involved in its consumption or creation, because they suffer significantly from something unrelated to the porn. Many I spoke with shared similar stories. Stories of broken families, deaths of loved ones, feelings of loneliness and isolation, self-harm, panic and anxiety attacks, and even eating disorders were common. That being acknowledged, it's vital to emphatically state that someone's suffering never gives them the right to inflict suffering on others. There are simply no exceptions to this.

Just as I have done with the celebrity porn deepfakers (and the Photoshop fakirs before them), it's also been my objective for this chapter to capture the perspective of a victim of deepfake pornography. I made several attempts to reach out to women – both famous and not – who have had their faces used in pornographic deepfakes without their consent. Of the ones who replied, most declined to talk about their experiences. As someone who has had traumatic physical experiences inflicted upon myself by others, I understand this.

While some try to reclaim their trauma, like Faraz Ansari, who now uses the Reface app on themselves, others take a different approach to tackling past trauma. Some write books to birth something good out of the bad, while others, for whatever reason, can just shrug it off, perhaps understanding that no single experience of suffering should be allowed to define their lives. In the end, there is no 'right' way to deal with trauma – and you don't always need to talk about it to overcome it.

One person who replied to my queries, however, was willing to speak about how she felt. She agreed to give a few short comments on her experience, provided her name remains confidential. She's 28 and works in finance. I was able to obtain the deepfake of her because it was posted to a public imageboard in an album with dozens of other photos of people who had been deepfaked. I believe the deepfake was created in an app that was based on DeepNude. Using her face in the deepfaked image, I was able to track down, via a simple reverse image search, the real person to whom the faked nude body 'belonged' in about five minutes – something in itself that should unnerve us all.

After I'd identified myself and told her about this book, she asked me to forward her the deepfaked image I'd found. She was already well aware of deepfakes and that people used the technology to create fake celebrity porn videos and funny YouTube videos. She told me she searched her social media posts for the photograph the person who deepfaked her used. After identifying it as a photo she posted to one of her public social media profiles, she believes one of her 'friends' may have created the deepfake.

'Honestly, I don't know whether to laugh or cry,' she tells me. 'And it's a little upsetting that one of my so-called friends is making this stuff of me. But given all the people who probably had access to this [original, non-nude] photo, who knows who it could be? It makes you think twice about posting to social media, though.'

I ask her what she would say to this social media 'friend' – if that's who it was – if she could confront them about the deep-fake they created without her consent.

'I know lots of lawyers, so I'd definitely scare the shit out of him by saying I'm suing,' she says. 'But honestly . . . I probably wouldn't. I mean, why bother? It's so much trouble to do that. But if I could confront him, I guess I would say I kind of get why an obsessed fan creates deepfake porn of celebrities . . . but one of me? An ordinary girl? A friend? What kind of loser does that? I mean, really, whoever did this of me is a fucking joke and they'll probably die alone if this is what they spend their time doing.'

Joke or not, her comments about suing the deepfaker bring up an interesting issue. Can you sue someone for deepfaking you? Is the creation of deepfake porn even a crime?

That's a set of questions with complicated answers, as we'll soon see. Pornography aside, though, there are ways deepfakes can be used to facilitate crime – crimes that may make others sit up and pay attention to the technology since the target of those crimes encompasses all genders. Crimes that make you realise just how vulnerable we all are – man or woman, young or old – to the many illicit and malicious uses of deepfake technology.

Chapter Four

Deepfakers for Hire

'Look, this is going to sound weird, but the deepfake I need you to make is a deepfake . . . of me.'

That's what I say on my first audio chat on an end-to-end encrypted secure messaging app with a deepfaker.

A deepfaker for hire.

A deepfaker for hire who will be known as 'Brad' in this book.

Like most deepfakers you find on less-than-reputable Discord servers and internet forums, Brad is only willing to talk to me because I've agreed to certain criteria. Criteria like giving him a fake name, not mentioning what server or message board we met on, and agreeing to move our conversation onto a secure messaging app, which, too, Brad says, must remain unnamed.

'OK . . .' Brad says.

I hesitate. I know this is going to sound so bizarre.

I let out a breath.

'And the deepfake of me,' I go on, 'it needs to show me committing a crime.'

Silence from Brad.

More silence.

'Uh, are you still there?'

'Yeah,' Brad says, as if just realising himself he's still there. 'What kind of crime?'

And really, I just want to stop because I know how batshit crazy this sounds, but . . . 'That's the thing . . . I don't want to decide what it is. I want you to pick the crime. Surprise me. Mugging. Carjacking. Just . . . I don't want to know what it is before you show me the final deepfake. I want to be surprised.'

More silence from Brad. And then a little breath of laughter.

'Are you really a journalist? Or is this some kind of setup?'

And even I have to admit, this *does* sound like a setup. But I reaffirm I'm really a journalist. And I remind him that when I reached out to him a few days ago, I included a link to a press release for this book and a bio about me.

'Yeah, but when you said you're looking for a deepfaker to make you a deepfake – even for this book – I thought you meant, you know, porn. But a deepfake of you committing a crime? How do I know you are who you say you are? How do I know you aren't some guy who wants to set this Michael Grothaus up by creating a deepfake of him doing something illegal?'

I'm relieved to hear Brad say this. It makes me think I've found a deepfaker who has a moral compass.

'And if you are setting him up, I could probably get in trouble for making the deepfake if he's not really you.'

Or at least a sense of self-preservation.

I'm quiet for a moment, thinking . . .

Two-factor authentication!

Two-factor authentication, or 2FA for short, is a security technique most people are probably aware of, even if they don't recognise the term. Two-factor authentication is that code you get texted by your bank or online service provider when you try to log into your account for the first time from a new device. Though you've successfully entered your username and password, the bank or service needs another way to confirm you are who you say you are. Banks and service providers do this by texting you that code, which you must enter on their site within a specific timeframe. Since they've previously verified your phone number, they know only the real you will get and input their code.

Two-factor authentication is one of the best user-facing security measures any bank or online service can provide. The problem is, Brad isn't a service or a bank, and our previous form of communication wouldn't prove I'm me. Thankfully, I'm a novelist and journalist – and I'm on Twitter.

'I'm— Michael Grothaus, the journalist who is writing this book is verified on Twitter,' I say.

Verified is that blue checkmark Twitter adds to certain accounts of public or media figures to confirm that the account under their name belongs to the actual person. That blue checkmark by @michaelgrothaus on Twitter, in other words, says that Twitter has verified that the person tweeting from that account is me.

'So using Twitter, I can prove to you that you're actually talking to me. Just – right now – give me a string of random letters or numbers – something that would be beyond the realm of possibility that Michael Grothaus, if he were not me, would tweet out at this exact moment.'

'I see where you're going,' Brad says.

I hear the soft clackity-clack of a keyboard. I assume he's navigating to my Twitter profile.

'OK,' he says, 'tweet this.'

He gives me a random string of characters to tweet.

I do.

'OK, tweeted,' I confirm. 'Do you see it?'

I hear a single soft key-clack.

'Yeah . . .'

'Cool.' I'm a bit relieved. 'And look, I don't blame you for being suspicious. I know someone contacting you to have you make a deepfake of themself must seem insane.'

'Oh, no,' Brad says. 'I've actually done that before.'

'You have?'

'Yeah. I did it for a guy last year. But I mean, it wasn't weird like what you want. He didn't want himself to be engaging in a crime.'

'What did he want to be doing?'

'He wanted to be the one fucking this Korean porn star in this one video. It was his favourite porn star and his favourite video of her. Blowjob, doggy, anal, ending with a facial – the usual type of porn video, but he just wanted to be the guy in it.'

'Jesus,' I say before I can stop myself.

Brad laughs.

'Yeah. To each their own. Was a long video, too. Like thirty-something minutes.'

'That must have taken a while to deepfake . . .'

'Not really. Training his face's model took about a day and a half, but that's no longer than most models take on my rig – if you want a good one, that is. Once that's done, DFL did the rest.'

By 'rig', Brad means his computer. If a well-trained model only takes a day and a half on it, I'm assuming Brad's rig has an incredible GPU.

I'm reluctant to ask because when I reached out to Brad initially, I told him that I couldn't pay him if he were willing to make a deepfake for me. It's a journalist thing – paying your sources is frowned upon. Still, I ask Brad how much the guy paid to have himself deepfaked into his favourite porno.

'That was a couple hundred.'

That's a bit on the pricey side. On many deepfake forums and Discord servers, you can find deepfakers willing to create custom deepfake porn videos for anywhere between $20 to $100. Two hundred dollars for a deepfake is the most expensive I've yet seen. However, I'm not going to say that out loud because Brad is doing mine for free. He's also clearly skilled, judging by some of his other deepfakes I saw before contacting him, which could raise his rates. Then again, 'skilled' is probably a misnomer when talking about a deepfaker. 'Technically proficient' and 'has great hardware' are probably better descriptors.

'So why a crime?' Brad asks. 'I get writing about having a deepfake made for the book might be interesting for people who don't know about the process, but why crime and not porn? And why of you?'

All good questions.

My original outline for this book didn't include narrating an attempt to get a deepfaker to make a deepfake for me. And it didn't even include me making a deepfake of myself as I could easily do in an app like Reface. But as I became more engrossed in my research, I knew I needed to show not only how easy it was for someone willing to take the time to learn the software tools to make a deepfake of the length and quality that your typical porn deepfakes are, but also to show how easy it is for someone who doesn't want to become a deepfaker to find someone to make a deepfake of that length and quality for them. In other words, I needed to jump into the world of deepfakers for hire. And if you want to hire someone to make you a deepfake, you at least need to tell them who you want deepfaked, and what situation you want that person involved in.

The obvious choice, of course, was hiring someone to make a pornographic deepfake. But I wouldn't have that created unless I could get consent from everyone who would appear in the final video (including the porn actress and any other porn stars in the video). And let's face it: even if you can prove you're writing a book about deepfakes, it's going to sound super weird contacting someone to ask, 'Can I have your permission to include you in a video of a sexual encounter you never took part in?'

Besides, after scurrying about in the digital underworld of deepfake porn communities for months as research for the previous chapter, I was ready to move on from the strictly pornographic aspects of deepfakes – and for a good reason. As we'll shortly see, deepfakes can be weaponised against a target in ways that don't always involve placing them in non-consensual porn – with equally tragic and devastating effects.

That's why I settled on a deepfake of me committing a crime – and a crime the deepfaker chose, not me. The reason I settled on this is because – whether it's Faraz Ansari and their Victoria's Secret Reface bullying deepfake, the 28-year-old finance professional's photographic deepfake or a Scarlett Johansson in a threesome deepfake, those victims all have one thing in common no matter what walk of life they come from: they aren't asked what type of deepfake they'd like to be placed into. They don't have that choice. So, though it's necessary that I reach out to have a deepfake made of me, I don't want to specify what type of deepfake it is other than it should be something I would find distasteful, shocking or horrifying to see myself engaged in.

I want to experience that visceral incredulity and dread of seeing myself unexpectedly doing something I know I didn't do.

'So that's why it's crime,' I explain to Brad. 'And it's why I need you to pick the scenario, not me.'

Brad is quiet for a moment.

'And so the base video—'

'I know it's a big ask,' I interrupt. 'But I can't help you pick it out.'

Brad takes this better than most probably would, especially since he's not getting paid for it. But, if there's one common trait I've found among deepfakers who hang out in online communities, it's that most seem to be fairly supportive of each other and are interested in propagating the craft. Many love experimenting with deepfake technology and they're willing to help others experiment too. They not only share their work, but their best practices, their constructive feedback, their face-sets and even their trained models. You don't see that type of comradery in creative groups of all stripes.

'I appreciate it,' I tell him, although the moment it comes out of my mouth, I realise how stupid that sounds. This is the weirdest conversation I've ever had.

'I'll need images of you – for the faceset. To train my model of you,' Brad says. 'Easiest is just to take a video of you. Set your phone up on a tripod and do a close-up on your face. Talk during it, so your face goes through a range of expressions. Also, move your head around every so often – side to side and such – so I get enough angles. Sixty to ninety seconds should give me enough images.'

I immediately understand what Brad means. Video is just made up of a bunch of still images, after all. Each second of video has almost 30 frames: 30 frames per second times 60 seconds is 1,800 images; times 90 seconds is 2,700. That's 1,800 to 2,700 still images of me Brad would get from a 60- to 90-second video of my face – more than enough to make a solid faceset to train a model of me.

But God, I feel like an asshole now.

'That's the other thing,' I add. 'I only want you to use a faceset extracted from media that's already existed online – something of me that's already been posted.'

I pause but start speaking before Brad has a chance to reply. I don't want him to think he's going to have to scour the web for images of me. Instead, I tell him that I have a video of me that's been posted online before.

'Will that work?' I say.

The reason I want Brad to use this video to source my faceset is I want it to be authentic to the way most deepfakes are created: the facesets of actual victims of deepfakes are usually sourced from the frames of an existing video of themselves that's already on the web. No porn deepfaker actually contacts a deepfake's target and gets them to send a pristine, purpose-made recording of their face to train the model with.

I'm relieved to hear Brad let out a small laugh. 'Can I at least get you to send me the video you are talking about?'

I tell him that's not a problem.

'One more thing,' he says. I hear some soft clacking on his end. Then he gives me a set of random characters. 'Can you tweet this?'

He wants me to do our version of 2FA again. These are different characters than before.

'Just to be safe,' he adds.

I do it.

'Thanks. You can never be too safe.'

After confirming some logistics, Brad says he'll be in touch with word on my deepfake's progress when he has the time.

Upon hanging up, I get the chills. The actual chills. Like a shiver runs down my spine. I've just given permission for a deepfaker – a complete stranger I 'met' in cyberspace – to make a deepfake of me *committing a crime*. Part of me is really interested to see what he comes up with – and how lifelike it will look; part of me is *terrified* to see what he'll come up with.

What would it be like seeing yourself in a video doing something you know you didn't do? Especially if it's potentially criminal?

This entire scenario, it's worth noting, is one of the biggest worries people have surrounding the threat of deepfakes. While deepfake technology being used to create non-consensual fake porn videos gets much of the headlines, the threat that individuals will be inserted into deepfakes showing them engaging in crimes they didn't commit can potentially be far more damaging.

Imagine someone with a vendetta against you. They want to inflict harm upon you – make you suffer. How do they do it? By creating a porn deepfake of you? Sure, that might humiliate you, but is that enough? What if they really want to hurt you – destroy your social and even legal standing? They can grab their smartphone and shoot a new video of someone else engulfing a parked car in gasoline and setting it alight. Then they take the custom video they shot and deepfake your face onto the arsonist's. Suddenly, you're the latest wanted person in the city.

Maybe that scenario sounds far-fetched, but it's perfectly technically achievable with existing deepfake technology. Yet, while, at the time of writing this, no one has been framed for a crime using deepfakes, other illicit uses of the technology (besides non-consensual fake pornography) have already been unleashed upon the world.

The first illicit use is closely related to framing someone for something they didn't do. But while framing someone is usually done to bring on legal problems, here deepfakes were used to humiliate and intimidate a person in order to change their behaviour.

While most deepfakers who create fake celebrity porn are careful to label their deepfakes as a parody for legal reasons, in 2018, investigative journalist Rana Ayyub didn't even get that courtesy. After an eight-year-old Kashmiri girl was raped in 2018 and the country's nationalist Bharatiya Janata Party (BJP) mobilised to support the person accused of the act, Ayyub was invited to speak on multiple media outlets about how some in India's political sphere were shielding the alleged attacker.

Within days of those media appearances, faked screenshots purporting to be tweets from Ayyub's verified Twitter account spread on social media.[58] Ayyub ostensibly proclaimed, 'I hate India and Indians!' and made other outlandish statements in the faked tweets. The following day, things got worse.

Much worse.

While in a café with a friend, Ayyub received a message from a source. 'Something is circulating around WhatsApp, I'm going to send it to you but promise me you won't feel upset,' the message read.

What was sent next was a video of Ayyub having sex. Well, it was her face on the body of a young woman having sex. It was a deepfake. And it was being shared among those in BJP political circles.

Ayyub started throwing up.

But things got even worse. The deepfake was then shared on a fanpage of the BJP leader. After that, it was shared 40,000 more times. *40,000.* Throughout this entire ordeal, strangers contacted Ayyub with messages stating, 'I never knew you had such a stunning body,' among other taunts.

This was pornographic deepfaking not for sexual gratification but targeted humiliation. And humiliation with the aim to muzzle a journalist. To get her to stop speaking truth to power.

'See, Rana, what we spread about you; this is what happens when you write lies about Modi and Hindus in India,' the administrator of a Facebook page that shared the video wrote to Ayyub.[59]

And this was back in 2018 when deepfake technology was far inferior to and less widespread than it is today.

Yet beyond targeted humiliation campaigns meant to get a person to change their behaviour, deepfakes can be used in other overtly criminal ways. Closely related to framing people for crimes and targeted humiliation campaigns, deepfakes can be used for blackmail and extortion.

The most obvious example of this is a scenario in which someone creates a deepfake of a target yet doesn't release the deepfake to the public. Instead, the deepfaker sends the target themself the deepfake. Perhaps it shows the target physically or sexually assaulting someone or engaging in taboo sex acts. The intent here is not to trick the target into thinking it's real – because they obviously know it's not. Rather the intent is to generate anxiety, fear and worry in the target that *other* people might think it's real should the deepfake be publicly released.

'Isn't it just much better,' the deepfaker will argue, 'to protect your reputation and prevent any legal inquiries by paying me money not to release this deepfake? A one-time payment of ten grand will make this all go away.'

Or perhaps the deepfaker isn't after financial extortion. Perhaps the deepfaker is after sexual extortion. Sexual extortion, or 'sextortion' as it's colloquially known, happens when someone threatens the victim with a potentially damaging scenario (public humiliation, loss of their marriage or social standing) unless they agree to gratify the extorter sexually – either in-person or digitally. Many forms of internet-based digital sextortion begin when the sextorter gets a victim to send them authentic nude selfies willingly, often by tricking the victim into thinking the perpetrator is a beautiful love interest. Once the sextorter has these authentic nude selfies, they then hold them over the victim's head: 'Send me one nude selfie a day from now on, or else I send the ones I already have to your boss/parents/spouse/children.'

With deepfakes, a sextortionist wouldn't need a victim to willingly send them that first nude selfie. They could just make a deepfake of the victim engaged in a sex act and force the victim to send them a real video of themself engaged in that sex act – or else they'll release the deepfake to the victim's family and co-workers.

Keep in mind that it's easy to assume you or I would stand up to a sextortionist attack such as this, but you never know what existing frame of mind a victim is in. Unfortunately, too many sextortion attempts have ended with the victim taking their own life rather than seeking help with the situation or risking public shame.

What's even more worrying about deepfake extortion rackets is that they can be automated. These take the form of theoretical ransomware software known as 'deepfake ransomware', or more succinctly the 'RansomFake', as some call it.[60] Regular ransomware is a malicious program that gets installed on your computer. Usually, such a program encrypts your hard drive, so you lose access to all your data. You'll only get your access back, the ransomware explains, if you click the link to pay the ransom.

Deepfake ransomware could work similarly. Instead of encrypting your data so you can't access it, this software first learns who you are from the data on your computer. It then finds publicly available images or videos of you online or on your computer itself and, finally, uses that media to make a deepfake of you in a compromising position. The ransomware would display the deepfake of you on your screen with a prominent timer counting down. If the timer reaches zero before the ransom has been paid, the deepfake of you gets released to the public.

The fact that deepfake ransomware could be automated means one bad actor with the technology could conceivably target tens of thousands of victims at a time – and even use the hijacked computers' processing powers to generate the deepfake. The more victims targeted, the better chance some will pay up, generating a hefty profit. And for those who do not? Well, enjoy seeing yourself with embarrassing household objects inserted into your orifices when you Google yourself in the future. Enjoy the rest of the world seeing it, too. This very real possibility is utterly chilling.

And we're just getting warmed up.

Leaving ransoms aside, there's also the worry of identity theft. So far, we've only talked about deepfakes being used to make people look to be engaged in something they never, in fact, did. In other words, the deepfaker takes your image to fake you onto someone else's body. Yet deepfake technology can be used for almost the exact opposite. That is, the deepfaker can take your image to fake you onto *his* actual body.

Why would a deepfaker want to make himself look like you? Mainly to fool biometric systems. The world is currently undergoing a shift to biometric authentication for security ver-ifications. And circumventing some facial recognition biometric authentication systems is small fry for an experienced deepfaker.

Already today, some smartphone banking apps and some dating apps will ask you to verify your identity by uploading a current selfie of yourself along with a government ID. This is another form of two-factor authentication. Deepfake software, however, can easily enable an attacker to generate a selfie of you on the fly, especially if that deepfaker already has a trained model of your face in his collection.

What's even more frightening than this scenario is deep-fakers can use a trained model with software that allows them to wear a digital mask of your face on theirs in real-time. This means the deepfaker could literally take part in a live video call wearing your face, and no one who didn't know the sound of your voice already would realise it wasn't you.

But what if the participants on the other end of the video call did know your voice? For example, say you were video conferencing with your mom or brother. Surely, they would recognise the faux you in the video call sounded different? Sorry, that little hiccup can already be easily overcome, too.

Remember, though the term 'deepfakes' is colloquially used to refer to fabricated videos created by artificial intelli-gence – that artificial intelligence is not just limited to syn-thesising faces out of thin air. The same AI techniques have long been able to create synthetic voices that can mimic an individual's actual voice. In this case, instead of the deepfake AI training on a faceset of photos of the target, the software is fed a small snippet of audio – the target's voice – which it then clones. This not only allows the deepfaker to make a target say anything they want them to in a falsified video

– like President Biden declaring he intends to confiscate all firearms. Such voice cloning deepfake software also allows the user to mimic the target's voice in real-time. That should have us all worried.

Why? Forget the fact that some financial institutions already verify a caller's identity by comparing their voice to a voice-print on file (and if they match, that's all the identification the institution needs to discuss the account with the caller – or have all the funds transferred), deepfaked audio impersonation is one of the few examples of non-pornographic deepfake crimes that have already been committed in the real world.

On a late Friday afternoon in March 2019, the managing director of a British energy company received an urgent phone call from Johannes, the company's CEO, who instructed his managing director to immediately wire $240,000 to the bank account of an energy supplier in Hungary to avoid being hit with late-payment fees. As the managing director had spoken directly with the CEO many times and recognised his voice, he paid the Hungarian supplier, as instructed, with the bank details provided.

It was only after the CEO made a second request that the managing director grew suspicious. He called the CEO directly. While on the call with the very confused CEO, the 'CEO' who'd contacted him called back a third time. 'The fake "Johannes" was demanding to speak to me while I was still on the phone to the real Johannes!' the managing director later lamented in an email.[61]

What had happened was thieves used deepfake software capable of cloning a person's speech in real-time and called the managing director pretending to be the CEO. As a representative of the insurer which later covered the company's $240,000 loss noted: 'The software was able to imitate the voice [of Johannes], and not only the voice: the tonality, the punctuation, the German accent.'

Now consider how easy it is for someone to get a recording of your voice – just a bit that's long enough to train deepfake software to mimic it – you may think it's hard to get a hold of such a recording, but it's not. It's really not.

Think about it.

All they need to do is call you up and get you to talk as they record the conversation without your knowledge. 'Oh, wrong number? Who is this then? But then why do I have this number? Can you repeat it to me, so I know I've dialled the right one? Thanks, and so sorry about that. You have a great night.'

Or if they're in a meeting or at a gathering with you, they simply need to take out their phone and use the built-in recording app to collect a sample of your speech without you noticing. Or they don't even need to be near you or call you to steal your voice. They could just rip your voice from a video you've previously posted to social media – even one posted years ago. All they need is a minute, maybe less, of your voice to train their deepfake audio tools to clone it, and then they can make you say whatever they want you to say, even in real-time.

Think about this the next time mom gives you a call. Is it really her?

There is arguably nothing more unique about us than the way we speak – our enunciations, how we pause, the rhythm of our syntax. Our speech is even more distinct than our face. (How often have we seen someone who looks exactly like someone else we know?) And considering that some thieves have already made off with almost a quarter of a million dollars by merely deepfaking a person's voice, imagine what other financial damage a deepfaker with grander, more illicit motives could do. Far from just scamming individuals, a deepfaker could use audio cloning technology to manipulate entire markets. What if a 'private call' between Apple CEO Tim Cook and friend and investor Warren Buffet got leaked to a tech rumours site? In this call, Cook, the man who built Apple into a $2.5 trillion company in under a decade of his leadership, reveals that he is about to announce his surprise resignation in twenty-four hours. The reason? Like Steve Jobs before him, he's been diagnosed with terminal cancer.

Apple's stock – and perhaps the entire market that day – would go into free-fall. Even if the call were quickly outed as a deepfake ruse, it's feasible that, within seconds of its original reporting, a quarter of a trillion dollars of Apple's valuation could vanish as institutional investors rushed to sell. And if the deepfaker who created the fake call of Cook shorted AAPL stock right before leaking the audio? Well, he could have made billions in those split seconds.

Let's pause and move past porn, intimidation, extortion and rackets: how else could deepfakes be used maliciously?

What about espionage?

For close to a decade now, it's been known that Chinese spies create fake profiles on popular social media platforms like Facebook and LinkedIn to befriend and connect with Westerners in positions of interest to the Chinese Communist Party.[62] These Westerners may work as defence contractors, researchers, civil servants, government officials or military officers. The spies set up these elaborate social media profiles complete with fabricated histories of their education, workplaces, interests and even families. They'll then complete their profile (on LinkedIn, for example) with a profile picture stolen from elsewhere online. Maybe they'll even populate their fake Facebook profile with pictures stolen from a Google image search showing them and their 'family' camping during a summer outing.

The goal of these fake online personas is to gather information about the people duped into accepting their connection requests. As many people are careless with the information they post on social media, a spy using this method can obtain not only information including a target's phone number and email address, but information about their children, lifestyle, and even personal and religious beliefs – anything the targets choose to post and regularly share on their private social media pages. The spies then use all this information to build profiles around these persons of interest – profiles that might lead to other attack vectors or help the CCP identify foreigners who may be willing to spy for them.

While creating fake online profiles is a clever ruse, once it was widely known China does this, it was surprisingly simple to identify which profiles were faked. You see, since the spies simply pulled photos of real people from the internet and used those as their profile pictures, all anyone had to do to determine if a Facebook friend invite or a LinkedIn connection request was from the person it appeared to be from was run the profile picture of the person who sent the invite through a simple reverse-image search. This is where you upload a photo to a search engine like Google or TinEye, and it tells you where that photo came from – such as from an existing social media profile or a stock photography company like Shutterstock. This is how I identified the 28-year-old finance worker who had a deepfake of herself posted to a public imageboard. If the reverse-image search reveals 'Tina Thompson', the 'marketing manager from Tacoma' who sent you the LinkedIn connection request turns out to be a photo of Deborah Alexander, a Vermont teacher, you know the profile is a fake. But this simple way of outing attempted espionage all changed once GAN technology came into being. You see, not only can GAN-based applications create forged faces of real individuals, they can also create realistic photos of completely fabricated people – that is, people who have never existed.

To see just how good GANs are at generating realistic photographs of people who have never walked this earth, you simply need to visit ThisPersonDoesNotExist.com.[63] The website was created in 2019 by Phillip Wang,[64] an Uber software engineer, who used a framework called StyleGAN[65] – which is graphics giant NVidia's bespoke GAN software. A visitor to ThisPersonDoesNotExist is met with a photograph of an individual. It might be a pretty 18-year-old woman, a 10-month-old baby or an 80-year-old man. Refreshing the page brings up an image of another person's face. Refresh the page again to see another person – and so on, and so on. You can keep looking at as many people as you want, marvelling at how though all faces have two eyes, a nose and a mouth, they can yet somehow look infinitely different. But, of course, the differences of the people in the images on this website aren't real. Nothing about the people in any of these images are. The only thing shared by all the individuals you see in these images is that they don't exist. They never have – and they never will. They are images of people created by GAN-based artificial intelligence every single time you click the refresh button.

None of the images exist until a visitor navigates to This-PersonDoesNotExist.com or refreshes the page. Only then – in that instant – does the GAN software actually create a new fake photograph of a new fake person. You can spend hours hitting your browser's refresh button on ThisPerson-DoesNotExist; hours looking at faces so real that if someone told you the people depicted never existed, you would think them insane. Yet every single photo generated in that split second of the refresh of ThisPersonDoesNotExist.com is nothing more than a construct of artificial intelligence.

It used to take gods to perform instant creation. Now, all it takes is GANs.

While that may freak you out, trust me, the spies probably love it. Thanks to GANs' ability to instantly generate photos of people who don't exist, Chinese spies are likely back in business on LinkedIn and Facebook, using fake profiles to connect with high-value targets. Doing a reverse-image search on a GAN photograph of a person who never existed won't return any positive results, after all, because that photo has never been posted anywhere else. It's simply been generated in milliseconds – with its only intended posting the fake social media profile meant to dupe foreign targets.

It's not just Chinese spies using GANs to create deep-fake photos of people out of thin air. Social media trolls use GAN-generated faces for the individual profile pictures of their thousands of bot accounts, scammers use it to catfish people in dating apps and Russian state operatives use it to flood social media with geotagged images of deepfaked people in order to dilute photos of political protests appearing in search results.[66]

I would be remiss if I didn't mention one final way deepfake technology can be used illicitly.

As we've seen: Yes, deepfakes can be used to swap people's faces into porn. Yes, it can be used to make a target think they're video conferencing with a particular person in real-time even though they are not. And yes, it can be used to create wholly fabricated people out of thin air.

All these malicious uses of deepfakes rely on artificial intelligence generating faces – real ones or not. But deepfake technology can be used to generate more than faces. It can be used to generate the very organs inside our bodies.

I know – what?

A popular target of hackers in recent years are medical facilities – hospitals, mainly. Hackers usually target these hospitals with ransomware – infecting computers and locking hospital staff out of patient records. While ransomware attacks on individuals can be financially and emotionally devastating, ransomware attacks on hospitals can be deadly. If a hacker encrypts patients' medical data, doctors can immediately lose access to vital information – like a patient's prior conditions, what medications they're on, and in what dosages – and they could possibly even lose access to real-time biometric data, like pulse and heart rate.

In 2019, researchers examined whether a hacker could use deepfake technology to generate realistic images of internal body organs – organs as they appear in CT and other 3D medical scans.[67] The researchers discovered that, yes, deepfake technology can be used to generate CT-grade images of internal organs that a hacker could then insert into a patient's medical records – replacing the patient's authentic CT scans with the deepfaked CT scans.

Now, why would a hacker do this? After all, if a hacker uses deepfaked CT scans to make it look like a patient has advanced stomach cancer, it's not like the patient actually gets stomach cancer. But the researchers noted several scenarios in which hackers might want to inject deepfake-generated medical scans into authentic medical records.

First, the hacker may want to inflict psychological or emotional distress on a specific individual. It's easy to see how someone being told their CT scan reveals they have cancer and will likely only live a few more months would find this quite traumatic. The researchers also noted that by leading someone to believe they have a severe health condition, they might be able to get that person to resign from professional life – such as a politician or CEO.

Second, the deepfake-generated medical scans may be used for monetary and competitive gains, such as sabotaging a competitor's research. And third – and most nefariously – by injecting deepfake-generated medical scans showing a severe condition when in actuality there is none, the hacker may lead a patient's doctors to believe that risky surgery is immediately needed to attempt to alleviate the problem – and by doing this, the surgery itself may put the patient in danger, or at least take them out of commission for a while.

Alternately, a hacker may use deepfake-generated medical scans to make a patient's condition appear much worse than it is – convincing doctors that surgery would be pointless. This may mean a patient with an operable disease in its early stages might forgo surgery that could correct the problem, believing the malignancy is too advanced for surgery to be effective – thus inadvertently allowing the operable condition to advance to a stage where surgery really would be ineffective.

Deepfake technology means we now live in a world where even our internal organs may not be as they appear.

In my research for this chapter, I was shocked by how many people offered deepfaking-for-hire services. The highest-profile celebrity porn deepfakers commonly offer such services in the online communities they frequent. It's also virtually impossible to be a member of a deepfake Discord server and not receive a private message out of the blue from someone advertising their deepfaking services.

However, given that the deepfakers advertising their services almost always refer to pornographic works, I wondered just how easy it would be to hire one of them to make a deepfake for a different illicit use – such as for one of the scenarios already described.

Thankfully, the answer seems to be it's not that easy. Most of the Discord deepfake channels I've managed to join post rules about what is and is not permitted. While some allow deepfakers to advertise their services, advertising for anything other than non-nude deepfakes like fancasts or 18-and-over celebrity porn deepfakes gets you kicked off the server. That means deepfakers can't advertise for making pornographic deepfakes of celebrities who are minors, minors or adults who are not celebrities (such as your college crush or a co-worker) and deepfakes that can be used in extortion or blackmail.

I also contacted several websites that advertise personal deepfake services for individuals, which you can find yourself by doing a quick Google search. Of the ones who replied, they specified their services are being offered for safe-for-work deepfakes. If you want Trump's face on a clown, they can do that. If you want to brighten your child's day by inserting them into a scene from *The Avengers* they will do that. If you want porn deepfakes, they said, look elsewhere, 'but you won't have far to look'.

And I hadn't looked too far before I found Brad.

We check in about five days after our initial audio chat.

'I got a hold of your novel.' He's talking about *Epiphany Jones*. 'Haven't read it yet but sounds kind of good.'

It's always nice when you get a compliment on your work from the guy you hired to deepfake you into criminal behaviour.

'I'm training your model now, by the way,' he goes on. 'I got a late start. The day job has been keeping me busy this week.'

By 'training', Brad is referring to the process in which the deepfake AI is playing its GAN game with my face. The software is making iteration upon iteration of me, each one becoming more lifelike. I know not to ask what his day job is. Instead, I ask how I'm looking.

'Eh, it's early still,' Brad murmurs. 'But it'll come out fine.'

I need to stop myself from asking what the destination clip is that he's decided upon, though I badly want to know, so I ask if he has a few minutes to talk about deepfaking for hire. I want to use his answers in the book.

'Ask away. I'll answer if I can. If I'm comfortable.'

When I ask, Brad says this is the first time he's ever worked on a non-porn deepfake for someone else. 'I've made non-porn deepfakes for me before. I've face-swapped actors into other movie roles just for fun – especially when I was learning.'

And how many times has he worked as a deepfaker for hire?

'Quite a few.'

How many is that?

'More than twenty, less than a hundred. How about that?'

Fair enough, I say. And it's all been celebrity porn deepfakes?

'All but the one guy who wanted his face inserted onto the male porn star's body in the Korean porn star's video.'

How much has he made in total from deepfaking for hire?

'I won't answer that. But it hasn't allowed me to buy a Rolex or anything.'

Have people asked him to deepfake non-celebrity women?

'More often than you would think.'

Who are these women prospective clients want deepfaked into a porn video?

'Most of the time, they don't tell me that. When they contact deepfakers, they usually say stuff like, "There's this girl I want a deepfake of,"' Brad says. 'When they phrase it like that, you know it's not a celebrity. They'd tell you her name if it was. But I've had some people – a few – straight-up say "my teacher" or "my friend".'

And you always decline requests for these 'ordinary' women?

'Yeah. I don't touch that.'

Why not?

Brad is silent for a moment. 'It's just a red line for me. Yours would be too, by the way – if you weren't the one requesting this of you. I just don't like doing real people in deepfakes – except, again, if it's them asking; like the guy and his Korean porn.'

'But . . . a celebrity is a real person.'

'You know what I mean. An actress . . . we don't see them as a real person, right?'

Brad explains we see them in make-up and costumes and lighting and with a fake name playing a fake character in a movie that is made up. 'We see *Star Wars*, and we fantasise about Princess Leia, not Carrie Fisher, right? That's what I mean when I say actresses don't feel real.'

He continues. 'With a real girl – someone's ordinary crush or whatever . . . look, you see a deepfake of an actress, and even if it's not watermarked, you know it's fake, right? You know Margot Robbie didn't skip off from Hollywood to go making skin flicks. You know that. But placing an ordinary girl into a porn? Even if you knew her in real life but didn't know about deepfakes, you could think, "Oh, I guess she's been in a porn."'

I sort of understand his internal logic, although I'm sure every celebrity would disagree.

I change the focus and tell him I've been researching all the ways deepfakes can be used nefariously. If he doesn't make deepfakes of ordinary women, I say I guess he wouldn't make a deepfake of an ordinary person for blackmail purposes?

'Hell no. Any deepfake I make will only be something that is obviously fantasy – no matter how real it looks. By the way, I should mention that I'm slapping a watermark on your deepfake – just in case.'

I understand. And I'm a little relieved. And, I do think Brad is telling the truth. It does sound like he is vehemently opposed to deepfaking any non-famous people without their consent. I ask him if he knows of any other deepfakers who have made pornographic deepfakes of ordinary women for clients.

'Without naming names, sure. Of course. Everyone has their red lines. But I will say that most deepfakers who focus on celebrity porn probably wouldn't waste their time on ordinary women – because what do they get out of it? What does the ordinary girl mean to them?'

Money?

'Nah, man. No one is getting rich making deepfakes.'

Tell that to the scammers who used deepfaked audio to make off with $240,000 in a few minutes.

'So out of all the deepfakers you know who do it for money,' I say, 'what percentage would deepfake an ordinary person without their consent?'

'Honestly, man … ten … no. Five per cent or something. Just five at most. Most deepfakers who do celebrity porn walk the line. They know an actress can't sue them for a deepfake. But a regular person probably could.'

Is Brad right? Are deepfakers who 'walk the line' and stick to creating pornographic deepfakes of celebrities really safe from legal action? What about other types of non-consensual deepfakes? Are there laws against that?

It's a topic frequently discussed among deepfake celebrity porn communities and, increasingly, in media and legal circles. And the answer is pretty complicated. The short version is there are a few scattered laws specifically addressing the use of deepfake technology against individuals – but some of those laws may not hold up in the end. Before we get into why, though, let's look at them – it won't take long.

In July 2019, Virginia passed the first-ever law against pornographic deepfakes.[68] They did this by amending a previous law that criminalised revenge porn – the sharing of actual sexual images of people without their consent. The 2019 amendment made the sharing of realistic-looking fake sex videos and photos without the consent of the person depicted in them illegal as well. Under the amendment, creating a non-consensual pornographic deepfake of someone is a criminal offence, though it is a misdemeanour and not a felony.

In September 2019, Texas became the first state to ban the use of deepfake technology to create videos of politicians within thirty days of an election.[69] The Texas law aims to help mitigate the threat that a deepfake could be made of a politician that could sway an election.

In October 2019, California followed Texas's lead by making it illegal to create a deepfake of a politician within sixty days of an election.[70] But California also passed another law that allows someone who is the victim of a non-consensual deepfake to sue its creator for damages.[71] It's important to note that, unlike the Virginia law, the California law does not criminalise the making of non-consensual pornographic deepfakes. It only makes it a tort – a civil offence.

Also in 2019, Maryland amended an existing law against child pornography making computer–generated images that are 'Indistinguishable from an actual and identifiable child'[72] illegal as well. Finally, in late 2020, New York State passed a law that makes it a civil offence to create a 'digitisation' of an individual in sexually explicit material who did not consent to appear in it.[73]

And that's it – six laws across only five out of the fifty US states, two of which might not even hold up. I'm talking about the Texas law banning the deepfaking of a politician thirty days before an election and the California law prohibiting the deepfaking of a politician sixty days before an election. Why? Because of First Amendment protections, the Constitutional guarantee that provides freedom of speech in the United States. This includes protections for the spoken and written word as well as visual content, such as artwork and, yes, pornography.

The First Amendment's freedom of speech clause not only gives one the right to express their thoughts without repercussions from the government, but it also protects forms of content some may find offensive or hurtful. For example, the First Amendment is what allows someone to burn the American flag. It's offensive to many, but it's usually legal if done safely, in a controlled environment, and the person burning the flag is the one who owns it. Likewise, the First Amendment is what allows people to openly criticise influential individuals in audiovisual forms, such as politicians in political cartoons or celebrities who are the butt of a late-night talk show host's opening routine. The First Amendment also protects most pornography. One of the only firm exceptions to this being pornography involving minors, which is always illegal. Yet if you want to make scat porn of Jesus Christ humping his way through the twelve apostles, while some people may find that grossly offensive, there's nothing illegal about it in the US because of the First Amendment's freedom of speech guarantees.

Of course, there are limits on freedom of speech in the US. You can't go into a crowded bar and yell 'fire!' if there isn't one because doing so could cause a stampede and injure people. You also can't spread harmful lies about someone and hide behind freedom of speech protections because the First Amendment does not shield defamers – something that we'll come back to in a moment. Likewise, you can't threaten to kill someone and claim First Amendment protections. The freedom of speech does not give you the right to terrorise others.

But since the First Amendment protects political speech and the criticising, parodying and satirising of public figures, it's unclear if the Texas and California laws banning the deepfaking of a politician in the runup to an election would stand up in court. If a judge were to rule those anti-deepfake laws violated a person's First Amendment rights, the laws would get tossed. But what of the Virginia law that makes it a criminal act to deepfake someone into a sexual act without their consent? And the California and New York laws that say someone who is deepfaked without their consent can sue for civil damages? It's unclear if those laws would stand up to a rigorous examination of First Amendment guarantees. This is especially true for pornographic deepfakes of actresses and other public figures, which, it is often argued, can be seen as satire or parody of a person of public interest, and thus the pornographic deepfake is protected speech.

What all this probably means is that pornographic deepfakers can take solace – for now – knowing that, under existing US laws, if they pornographically deepfake a famous person, unless that person is under the age of 18, the First Amendment may protect them: many deepfakers believe it does. This is why most pornographic deepfakers stick to fake celebrity porn and watermark or title the videos as fake creations or parody. It gives them as much legal protection – or at least, legal arguments – as possible. They can say their deepfake is protected speech and not defamation or harassment because it concerns a public figure, and no right-thinking individual would think the deepfake was real due to its title or watermark identifying it as a fabricated video.

But what about in jurisdictions with no First Amendment protections? After all, the United States is only 1 of 195 nations. Could a celebrity sue a deepfaker if a deepfaked porn of them is posted online for all to see in India, China, Germany or Bolivia, for example? Possibly in the future – if those other countries pass anti-celebrity porn deepfake laws. However, even then, just because it would be possible to sue a deepfaker, it doesn't necessarily mean a celebrity would. The reason, once again, as with so many things in life, comes down to mathematics.

There are thousands of Photoshopped nude pictures of Scarlett Johansson, for example, and hundreds of pornographic deepfakes of her – and more crop up each week. These fake porn images and videos are made by hundreds of creators – each of whom would need to be sued individually. One lawsuit on its own is expensive and time-consuming. But hundreds? Even a megastar like Johansson may eventually run out of the money needed to fund so many suits in jurisdictions all over the world. And that's if her legal teams could even hunt down and identify those making the glut of her non-consensual fake porn – an almost impossible task in itself. Johansson herself has said, 'The fact is that trying to protect yourself from the internet and its depravity is basically a lost cause,' noting that she believes options to fight fake celebrity pornography are 'a useless pursuit, legally, mostly because the internet is a vast wormhole of darkness that eats itself'.[74]

Even discounting non-consensual pornographic deepfakes, hers may be the most apt description of the internet ever spoken.

Yet while celebrity porn deepfakers may rest comfortable knowing that – so far – First Amendment protections in the US and a lack of anti-deepfake porn laws around the world may guarantee their safety to deepfake celebrities without repercussions, they should know that when it comes to deepfaking non-celebrities, those same protections may not apply. That's because while most countries do not have any type of laws explicitly forbidding the deepfaking of people without their consent, existing, tangibly unrelated laws can mean the act of non-consensual deepfaking is already illegal.

This illegality has nothing to do with deepfake technology, per se, but rather how that technology is used. If a non-consensual deepfake is made of someone without their permission and the victim knows who did it and have proof of it, they can bring civil proceedings against the deepfaker or, in some situations, ask that criminal proceedings be raised against the deepfaker by asserting the violation of laws not related to deepfake technology.[75]

For example, if a deepfaker creates a deepfake of a person with the intention to extort or blackmail them, existing civil and criminal laws against extortion and blackmail would stand. In other words, a deepfaker can't argue that since deepfaking itself is not illegal, no crime was committed. Using a hammer isn't illegal in most instances, after all, but it is if you're using it to smash in someone's windshield.

Likewise, while a deepfaker could argue that his creation of a non-consensual pornographic deepfake of an 'ordinary' person is covered by First Amendment protections, the person depicted in the deepfake may still be able to file suit against the deepfaker under existing harassment, false light or defamation laws. These laws relate to the intentional personal, emotional or reputational injury of another, which aren't protected by free speech guarantees in most instances.

The legal avenues for bringing suit against a deepfake creator don't end there though. Creating a non-consensual deepfake (pornographic or not) could leave the deepfaker open to suit under Intentional Infliction of Emotional Distress (IIED) laws. These laws are useful when harassment can't be proved, but the victim can prove the deepfaker willfully sought to cause them severe emotional distress.

Harder to prove, yet still a possible avenue for bringing suit against a deepfaker, is right of publicity claims in which the victim could argue the commercial value of their personae was violated. Understandably, this claim would hold up better for celebrities than non-famous persons – though a celebrity exploring this avenue in the United States could see it nullified thanks to the aforementioned First Amendment protections allowing satire and parody of public figures.

But remember that in a non-consensual pornographic deepfake, there are always at least two victims – even if the deepfake is of a solo sex scene. The primary victim is the person whose likeness – whose face – is deepfaked into the video. But the secondary person who's been victimised is the original porn performer, who likely never gave their consent to have their body used with another person's face.

Could that porn performer file suit against a deepfaker? Perhaps – but harassment, false light, or defamation avenues would likely be out. The performer would also have a hard time arguing Intentional Infliction of Emotional Distress. However, if the performer owns the copyright to the original porn video, they may be able to hold the deepfaker liable for copyright infringement.

Of course, the ironic thing is that while non-famous people arguably have more and better legal options for bringing claims against a deepfaker than celebrities do, they are also much more likely to lack the means to do so. Harassment, false light, defamation and IIED suits can all be challenging to prove and the process drawn-out and thus very costly.

That's where a final set of existing laws could come in: laws regarding revenge porn – the violation of which is easier to prove. These are laws that forbid a person from sharing intimate images of another, even if those images were consented to when they were recorded. Revenge porn laws are designed to stop jilted exes from publicly humiliating past lovers by spreading nude photos of them all over the web.

In the United States, forty-six out of fifty states have some form of revenge porn laws on the books. Only two – Virginia and California – have specifically added terminology that states it is also criminal or tortious, respectively, to share non-consensual pornographic imagery that has been artificially created. But that doesn't necessarily mean that existing revenge porn laws in other states wouldn't cover non-consensual pornographic deepfakes, too, even if artificial media isn't specifically mentioned in their statutes.

Regardless, these existing revenge porn laws make it easy for states (and other countries) to simply and quickly amend existing legislation to make the sharing of AI-created non-consensual pornography of another person illegal *without* making deepfake technology itself, or the creation of other types of deepfake media, illegal. You'll understand why this is so important in the next chapter.

But first . . .

It's a sunny afternoon in what looks to be somewhere in South America. A tourist rides his bicycle in a bike lane that runs along the shoulder of the main road. He's following a group of other cyclists. They're all part of a tour group exploring the city. The tourist approaches an intersection. On the other side of the road is a red building and, in front of that, a T-junction. The crossroad and the main road's intersection is denoted by two pedestrian crossings, one on either side. The tourist cycling passes over the first one as a red van makes a right turn from the crossroad onto the main road and drives past the tourist in the opposite direction.

We're seeing all this from the vantage point of the tourist. He's recording his bicycle tour with a head-mounted GoPro camera.

After that van makes its right and passes the tourist, from the same crossroad comes a white Honda motorbike. But instead of turning onto the main road as the van did, it cuts right across it, breaking barely in time to stop from getting hit by a minivan laying on its horn. As that minivan passes, the tourist's GoPro feed captures the Honda again, now screeching to a stop on the second crossing – right in front of the tourist's path.

'Whoa, *amigo!*' the tourist says to the Honda motorcyclist now blocking the bike lane.

This motorcyclist on the Honda is wearing white trainers and blue jeans, with a black baseball jacket and a scarf around his neck.

He's also wearing my face.

The tourist with the GoPro on his head, he scoots his bicycle around me on the Honda and continues down the bike lane.

'Jesus Christ,' the tourist mumbles at the near miss.

Seconds later, the footage shows, I come riding up alongside the tourist again as he's cycling.

And I'm pretty impressed with myself, actually. I've only ridden a motorcycle once in my life. I was seventeen, and after driving just five feet, the motorcycle promptly tipped over. But now, as the video shows, I'm expertly driving a Honda – one-handed, no less – on a narrow bike lane, right next to the tourist cycling down it. And I'm riding the Honda with one hand because my other hand is reaching into my waistband – to pull out a gun.

I – the me in the video – mumble something . . . in Spanish.

'*Give me the backpack!*'

The tourist, the one recording all this with his head-mounted GoPro, stops his cycling dead cold as I, Michael Grothaus, am now pointing the gun – a revolver, it seems – right at him.

I speak rapidly in Spanish as the tourist responds, 'Whoa! *Amigo! Amigo!*'

He tries reversing on his bicycle, pushing back with his feet. I do the same on my Honda, all the while gripping my revolver in my right hand as my left holds onto the steering.

In Spanish, I keep ordering the tourist to do something.

'I don't know what you're saying, man,' the tourist insists.

If only he spoke Spanish like me.

But I'm not having it. My mouth bends into a firm line, and I repeat my order in Spanish as I thrust the revolver at him again.

'The bike?' he guesses and hurriedly hops off it and backs away.

But the bike isn't what I'm after, dammit!

I get off my Honda.

That's when the tourist bolts.

The GoPro on his head shakes as he sprints down the bike lane, away from me and my revolver. But he's not safe.

Turning his shaky head reveals I'm in hot pursuit, revolver gripped in my hand. Within seconds I overcome him, jumping right into his path. Now I'm in the dead centre of his GoPro's shot. It's a medium close-up – an MCU for you filmmakers out there. The shot shows my chest up to my head.

Staring myself squarely in my face . . . it *is* undoubtedly me, one hundred per cent. I mean, not only would my mom recognise me, but so would someone who'd only met me a few times. There can be no mistake – this is Michael Grothaus.

Yet I have an expression on my face I've never worn before. It's one of extreme agitation. No. More than that. It's one of complete desperation.

And once again, I miraculously speak Spanish, demanding something of the tourist.

'*Amigo* . . .' he pleads.

I'm not his *amigo*. He's not listening to me. This angers me. It makes me even more anxious. I cast a glance to see if anyone is coming to help. But no, no one is. And so, right beyond the edge of the frame, I thrust the revolver again, directly into his body.

'*Amigo!*' the tourist begs as the barrel prods him, my finger on the trigger. '*Amigo!*'

'Jesus Christ,' I – the real me – think as I watch myself violently assault a man in broad daylight. 'Just give me the fucking backpack. It's not worth dying over.'

The revolver's going to go off.

I'm getting too agitated.

My finger is going to slip, and this tourist is going to die for his fucking backpack.

My – the real me – my throat is dry.

I know I'm going to kill this man. I'm just waiting to hear the bang – the sound that rings out as I take this tourist's life for a sack of cloth and whatever's inside.

And that's when I hear it – both the real me watching the video and the me in the video robbing the tourist at gunpoint.

'*No! No, amigo! No!*' someone shouts.

The tourist with the GoPro on his head turns, and the footage reveals a bystander has rushed to his side.

I – the me in the video – I put my revolver back in my jeans pocket, saying something I don't understand, and dart out of frame. That's when the tourist with the GoPro on his head and the bystander who came to assist him sprint down the bike lane in the opposite direction like their asses are on fire.

'*Run! Run! Run! Run!*'

It's many moments before the tourist with the GoPro on his head dares to look back.

Am I in pursuit?

No – apparently not. I've jumped on my Honda and sped away – my assault and attempted armed robbery a failure.

Where did I ride off to?

Nowhere – technically, anyway. Because I was never there.

I was never there speaking Spanish, riding a Honda one-handed, while violently shoving a revolver at another human being. Yet despite those facts, a video now exists that shows I was.

'So, what do you think?' Brad asks. We're on an encrypted audio chat. It's three days after I last spoke to him. *Only* three days.

We've been on the chat for less than five minutes. It started by him giving me a URL to a private server where he's uploaded the deepfake for me to stream. The entire video itself – the first deepfake of me I've ever seen – is just ninety seconds long.

Just ninety seconds – but each felt so much longer.

What do I think?

What do I think?

This isn't the deepfake I was expecting. When I told Brad I couldn't help him find a destination video, because I wanted to be surprised, Brad noted he'd probably end up asking a friend to fake beat him up and film that. He even joked he had a pair of nunchucks he could use in the stunt. So, I assumed at worst I'd just be watching a short generic clip shot on someone's phone – a few badly choreographed punches thrown – and my face on the attacker.

I didn't expect something so . . . real.

Really, I feel a bit sick.

I already told you that I was bullied in high school. Unfortunately for me, the bullying started much earlier. I was bullied relentlessly in grade school, too. And in some ways, the grade school bullying was worse. Why? Younger kids usually have better imaginations than teenagers. They're more creative. When kids bully another kid, they often make up stories to get the subject of their wrath in trouble. Lying about someone else is one of the first mean things a kid learns how to do.

The bullies tell the teacher you stole the apple from her desk. They tell the teacher you were the one calling them names. They tell the teacher they caught you cheating. They make people believe you did something you really didn't do. The lies my bullies told about me were the first time I realised words have power. Words *are* power. And as that's the case with words, imagine how much power video has.

Now, seeing this deepfaked video of me – showing me in a realistic scenario willing to inflict physical harm on another person, well, it feels like the literal manifestation of one of the worst fears any person could have. Someone who watched this could believe I was an armed robber. And I know I asked to be deepfaked, but I never fathomed it'd feel this horrible seeing myself do something so terrible to someone else *that I never really did*.

'Hey,' I say after staring at the frozen frame of my deepfake for another moment. 'There's no watermark on this. You said you were going to watermark it.'

Watermarks are a clear sign a deepfake is a deepfake. Without them . . .

'Yeah, honestly, I forgot,' Brad says. 'I only got done with it a few hours ago.'

I nod, not that he can see that.

Brad tells me it took several days to train my model. That's where the artificial intelligence underpinning deepfake software learns how to forge my face. 'And I gotta say, this might be my best deepfake yet,' he says. 'The celebs I've done have nothing on yours.'

And I gotta say, that doesn't make me feel better.

Brad is right to brag though. It *really* looks like me. Well, the parts I'm in. Though the clip is about a minute and a half, the deepfaked portions featuring my face total less than half that. And while my likeness is clear, it's not perfect. I notice a few flaws. One is when the real armed robber quickly turns his head. During that action, my face doesn't blend perfectly. Another is that the real armed robber's hands were more tanned than my deepfaked face. Brad notes he corrected the deepfake to make my face look a little darker, but he probably should have done more.

I dismiss his self-criticism. 'No—I mean, look: you did a really good job. It's so good it freaks me out a little.' That's a lie. It freaks me out a lot.

'I thought it might,' Brad says.

I'm quiet for a moment. 'You can now make more of me, can't you?'

Brad knows what I mean. 'Yeah. Now I have a trained model of your face, so it would be easy, or much easier, to insert you into other videos.'

I scrub through the embedded video in the browser again – my deepfake. I see myself riding up on the Honda. I see myself pulling out the revolver. I see myself with that look of sheer desperation etched onto my face.

I remember the sound of the fear in the tourist's voice as I thrust the gun into his side.

'*Amigo! Amigo!*'

Why does the Michael Grothaus in this video want to harm this tourist? Does he want his backpack and its contents for his own use? To sell for drugs? To sell to get food for his family?

Looking at myself, I can't help but pity this alternate me.

'I'm sorry your life has led you to this,' I want to tell the me in the video.

Thankfully, the Michael Grothaus in this deepfake doesn't exist in reality. He isn't suffering somewhere out there. He can't be harmed or harm others. The real me, though? This video could harm the real me if people were to see it and didn't know it was a deepfake.

If there's any consolation for the real me, it's that the source video Brad extracted my faceset from is old. That means I look younger in it. Also, my hair has never looked like it did in the video. That means if Brad goes off the rails and starts pumping out deepfakes of me committing more crimes, I could at least note the discrepancy between the younger me in the deepfake and how I look in real life now. And in Brad's deepfake, there's also the fact that I'm speaking fluent Spanish, which I can't. Being able to speak Spanish and ride a motorcycle one-handed sounds so cool – and are two ways in which this fake Michael Grothaus one-ups the real me – but my lack of either of those talents is a relief right now.

I scrub the video again. And again.

'*Give me the backpack!*'

And, still, I feel a bit sick.

But I want to remember this. I need to. And you, reader, I need you to remember this, too. I need you to remember how horrible I feel *despite being the one who asked to be deepfaked.*

Now, remember all those women all over the world being deepfaked. Remember, *they never even got that choice.* But there's something else they didn't necessarily get from many of us, too: a real effort to understand what they've gone through.

Since beginning work on this book I've had a number of discussions with friends both male and female about deep-fakes – specifically, their non-consensual pornographic aspects. When I talked to my female friends about it and asked them how they would feel if they found a pornographic deepfake of themselves most, not all, but most replied with descriptions such as 'scared', 'horrified' and 'humiliated'.

When I asked my male friends that same question, most, again, not all, but most, answered along the lines of 'I honestly wouldn't care that much' and 'I guess I'd laugh it off'.

These are two *vastly* different types of answers to the same question.

This is despite both the men and women I talked to being in total agreement that creating non-consensual porn of another person is wrong – not one of them disputes that. Yet because of how the spectre of non-consensual deepfake pornography makes us *feel* we see its threat potential very differently.

Now this discrepancy between how males and females answered the *feel* question in the way they did probably has a lot to do with gender power dynamics and imbalances in society at large as well as the fact that pornography has, historically, been created and consumed with the male in mind – men usually have most, if not all, of the power and control in a pornographic video, thus it is less threatening and humiliating to them, which makes it easy for men to dismiss it as relatively harmless.

Yet, another factor that likely contributes to this discrepancy lies in the fact that we – people, regardless of gender – are usually just pretty bad at putting ourselves in someone else's shoes. We see the world through our eyes – our experiences, our fears – not someone else's. I won't pretend I know how to change that, but it's something to keep in mind the next time it's thought someone is exaggerating about the threat non-consensual deepfake pornography poses: you may be seeing it through your eyes when you should be looking at the problem through theirs.

I will also not pretend to know exactly what it's like to feel what a woman who has been the victim of a pornographic deepfake feels. I would be lying if I said I could. But after

seeing myself deepfaked into a situation that's even more unexpected and visceral than what I'd braced myself for, I now
have a better conceptual understanding of what people feel
who've been deepfaked into situations that are unexpected and
upsetting to them.

That's not to say that in all the years I researched fake celebrity pornography I lacked compassion for the film stars –
these people – these flesh and blood humans with feelings of
their own. But any compassion had always taken the form of
a quiet sympathy. I like to tell people's stories – whether the
stories of characters in my fiction or the stories of characters
in real-life – but I don't like to pass judgement on any of
these characters. I feel that's the job of the reader – if they so
choose – not mine. So as a journalist, and even as a novelist,
I've always tried to remain impartial to, or at least tried to
withhold expressing my personal judgement of, the people I
write about, including the ones who create or consume fake
celebrity porn.

But now . . . look, you can never truly know what
someone feels until you experience their suffering yourself.
You can't really know what it *feels* like to be blamed for
something you didn't do until it happens to you. You can't
know what it *feels* like to lose a parent until it happens to
you. You can't know what it *feels* like knowing your death
is imminent until that imminence is upon you. You can try
to put yourself into the person's shoes, of course, to run
through a well-intentioned intellectual exercise to attempt
to understand what they are experiencing – but as I've said,
we're too often so bad at doing that. And besides, until it
actually *happens* to you, you can't know what someone else's
suffering is *really* like. Until it happens to you – at best – you

can only feel *for* them. You can only *sympathise*. You can only feel *with* them – you can only *empathise*, in other words – when something similar happens to you, too – when you've suffered in the same way they have.

Now, because of this deepfake – *my* deepfake – I'm closer to feeling *with*. Of being more capable of putting myself in someone else's shoes. Though, of course, my experience still isn't completely the same, is it? At least I had the opportunity to consent to being put into my deepfake and brace myself as best I could for whatever it might show. It's an opportunity others haven't gotten and won't get in the future.

'Doesn't this shit scare you?' I say to Brad, giving the slider below the deepfake of me one last scrub.

'Honestly?'

No. Lie to me.

'I don't know,' he goes on. 'New technology always scares people. But I have my red lines. Only obviously fake scenarios. But yeah, I mean, people could make some messed up stuff if they wanted. But it's all dependent on those red lines. Most deepfakers have them. What you gotta worry about is the ones who don't.'

I understand that more now than ever.

'You sure you want me to destroy it?' Brad asks as we're wrapping up our call. He's told me the reason he sent me a link to a non-downloadable embed on his web server is that metadata in any raw file could, potentially, link it to its creator. He's told me I can make a screen recording of his deepfake of me as long as I leave the URL in the browser out.

But I don't want it. And I don't ever want to be tempted to show it to someone. Besides, part of the agreement I made with Brad was I won't show his work to anyone.

You can never be too safe.

'And you'll delete my training model and everything else, too?'

'A promise is a promise,' Brad responds. 'And no offence to you, but your face doesn't have anything on—' He names the celeb he's deepfaked the most. 'My type is unobtainable movie stars, not journalists.' He laughs.

'Well, thanks for all this,' I say to the guy who just created a criminal deepfake of me, and who literally has it in his power to create more if he's lying that he'll destroy the source video, my faceset, and my training model.

Yet something in me believes him – despite his questionable pastime. He's affable in a way that's similar to Skitz4twenty.

'Oh, hey,' he says before we say goodbye. 'I'm on chapter five of your book.'

He's talking about *Epiphany Jones*.

'Funny stuff, man. Never thought fake celebrity porn would make it into a novel. Who knew?'

And I never thought I'd be mugging a guy at gunpoint while speaking Spanish and driving a motorcycle one-handed.

Who knew?

A day after I said goodbye to Brad, I kept thinking about my deepfake. Not only about the Michael Grothaus in it with the horrible life doing horrible things, but about the actual video – the real event itself. I never asked Brad where he found it. He got my faceset from a personal video I provided him, but I don't know where on the internet this destination video came from. I don't know if there's a larger context of the real story to be discovered.

It takes me less than ten seconds to find the actual original video with a simple Google search. It's a video from years ago and is the record of an event that happened in Buenos Aires, Argentina.[76] The full video is just over 2 minutes and 40 seconds long. The good news: the tourist who was assaulted regrouped with others in his bike tour. Even better: they were all safe. After regrouping, they found a local police officer.

As for the armed robber – he looks nothing like me. Yet I still feel for him when I see that agitation on his face, that complete desperation. It's an experience we'll have always shared. In that way, we are the same.

As I'm writing the above words, I receive a ping in a Discord server. It's a direct message from someone I've never heard of.

'Big sale! 3 fake vidz,' it reads. 'One actress each.'

'All 3 deepfakes up to 10 minutes each.'

'$40 total.'

'7-day turnaround.'

'Sale ends Friday.'

Chapter Five

Deepfaking for Likes and Profit

The Walt Disney Company is one of the most beloved companies on the planet, and it's one of those rare companies that will soon see its centenary. Originally founded on 16 October 1923, by its namesake Walt Disney, the company spent the better part of the twentieth century churning out beloved cartoon characters, animated films destined to become classics and theme parks that can make even the oldest adult feel like a kid again.

And that was only during its first fifty years of existence. During the second fifty years, Disney, the brand, only became more embedded in our lives – in everything from our classrooms to our homes to our very consciousness. As children, we grew up with Mickey and Minnie Mouse backpacks we took to school every day, rushed home to watch our favourite tween shows on The Disney Channel, and then, on weekends, begged our parents to take us to the latest Pixar (another Disney brand) film.

As adults, we love sharing these same childhood experiences with our own children. We get as much joy as they do when we take them to Disneyland. We feel the excitement they feel when we settle onto the couch with them and flick through Disney Plus to introduce them to *Star Wars* (now Disney-owned), which thrilled our imaginations as kids all those years ago. And sure, we say we're watching a movie like Pixar's *Toy Story 4* or Marvel's (another Disney brand now) *Iron Man* because we know our kids will enjoy it. But admit it – it's as much for us as them.

The 'it's for the kids . . . but the adults, too' mentality is what, above all, defines Disney for what it's known best: being the world's most prominent *family* brand. When you consume a Disney product, you know what you're going to get ahead of time. A Disney product is wholesome. A Disney product is safe. A Disney product is appropriate for the ones you love most in your life.

Disney *is* family values. And for this fact, Disney has been richly rewarded. And I mean *richly*.

By 2019, The Walt Disney Company boasted ownership over eight of the ten highest-grossing films of all-time.[77] If you expand that to the top twenty films of all time, Disney owns fifteen of them. In fiscal year 2019, Disney pulled in $11.1 billion from the box office alone[78] thanks to its family-friendly titles. That same year its theme parks, experiences and consumer products generated another $26 billion.[79] 2019 also saw the Disney Plus streaming service launch, which by March 2021 had 100 million subscribers,[80] bringing the company close to $800 million in monthly revenue.

Is it any wonder then that in March 2021, The Walt Disney Company was worth over $350 billion? That's over one-third of a *trillion* dollars – a market cap the company will likely top in the next few decades. And they've accomplished all this over the last century thanks to content and messaging reinforcing the company's image as one that is the world's leading, wholesome family brand.

Given Disney's carefully crafted image – and how important it is to the company's bottom line – you may be surprised to find then that when, in 2017, the New York State Assembly introduced a bill to make non-consensual pornographic deepfakes illegal in the state, The Walt Disney Company vehemently opposed it.

Yes, Disney, the House of Mouse, the world's most prominent family friendly brand, the company that owns The Disney Channel – many of whose former starlets like Miley Cyrus, Selena Gomez, Vanessa Hudgens and Zendaya are now among the most pornographically deepfaked women in the world – came out to oppose a bill, one of the first ever, that would have made the non-consensual pornographic deepfaking of both celebrities and 'regular' people illegal.

But why on earth would the ultimate family friendly brand do this? To help you understand, I want to introduce you to Aliona Pole.

At first glance, browsing Aliona's social media profiles makes it appear she is no different than most 18-year-old women. Her Instagram feed is populated with the more curated kinds of images young people tend to post on the platform: close-ups showing her freckle-faced and stylish, pixie-cut cropped strawberry blonde hair. Then there's the obligatory lunch post – an overhead of Aliona holding a delicious veggie burger in her hands, ready to bite in. Scroll further and you'll see Aliona on the subway, then a shot of her looking at herself in the mirror, perhaps contemplating life or just the beauty of youth. Scroll more and you'll see Aliona in front of a Christmas tree outdoors, gripping a warm red takeaway cup of coffee.

Her TikTok feed, on the other hand, reveals a less curated Aliona. The short videos she posts there show her more relaxed – and playful. One video makes it appear as if she's reading a book as she reclines in a seat against a plane's window. But as the camera pulls back, we see it's a fake-out – Aliona is just messing with her audience. In reality, she's in her apartment, reclining against the washing machine, the door of which, when framed in closeup, looks like the porthole windows found on some airplanes. Another video shows Aliona working on her Mac-Book. A third shows her trying to exercise on a yoga mat only for her dog to interrupt, licking at her face.

In a recent interview, Aliona said she's concerned about the same issues that many people her age are: the environment, rational consumption and the effect of digital life on her psyche. In Aliona's eyes, you should only consume what you need, look after the earth and, despite her social media presence, she feels it's essential to maintain good digital hygiene. By that, she means she recommends being selective about what online information you consume, think critically about it before acting upon it, and most importantly, entirely get away from digital devices like your smartphone every so often to rejuvenate your mind and reconnect with those physically around you.

Yet, despite all this, Aliona is not like other people her age. Far from it. For starters, she's massively more successful than the average 18-year-old. Aliona is both a fashion designer and a model – and though she's been in the business for less than three years, she's already had more success than most twice her age. And 2020 was a particularly banner year for Aliona. In March, she modelled the latest fashions of celebrated Russian designer Alena Akhmadullina for *Vogue* magazine. A month later, she modelled FKSHM's new collection of eyewear with interchangeable lenses. Then in June, she launched a collection of clothing with AliExpress at the Mercedes-Benz Fashion Week in Russia. AliExpress is one of the biggest online retailers globally and is owned by China's powerful Alibaba Group. Not just anyone can collaborate with them.

So yeah, while most of us were stuck in some form of lockdown during 2020, Aliona was out there having experiences and successes few could ever dream of. Yet this isn't the only reason Aliona Pole is not like other people her age, despite the homey familiarity of her TikTok videos and her more curated life on Instagram.

The other reason Aliona isn't like other people her age?

She doesn't exist.

Despite working with some of the most prominent brands and designers in the world, she's never once breathed air. Despite her dog playfully licking her face while she tried to work out, she's never once felt the touch of its fur. And yep, despite her IG pic of that yummy-looking veggie burger, she's never once tasted the deliciousness of its grilled mushrooms.

You see, Aliona is a digital human. Or, more precisely, a virtual influencer. She exists digitally in cyberspace and nowhere else. Yet despite this, she does get those prestigious real-world sponsorship opportunities just like some human social media influencers do. And that interview she gave? It was with *Virtual-Humans*[81] – a site that tracks the industry.

It may be surprising to you but Aliona isn't the first virtual influencer, and she certainly won't be the last.

It used to be that celebrities, including actors and athletes, were the most influential spokesperson star-power any brand could have. Want to sell espresso machines in Japan? Get George Clooney for the ads. Athletic shoes in America? Michael Jordan is your man. Perfume in Paris? Natalie Portman and Scarlet Johansson will send your sales soaring. But in the early 2010s, a new breed of marketing spokesperson appeared: the social media influencer. At first, these social media influencers mainly encompassed fairly well-known figures already: reality stars, models, and musicians – famous people who used social media to expand their mindshare reach. But as apps like Instagram grew ever more popular, soon, 'ordinary' people started racking up followers in the millions, rivalling or even outdoing the follower counts of those reality stars, models, and musicians.

Brands quickly sat up and took notice of this. Why? To many brands, a plain old so-called 'social media influencer' posting from their bedroom in Boise with a million-follower count was as good as an old-school celebrity with a million followers. After all, either way, there are a million pairs of eyes looking at virtually every photo they post. Now, if only that social media influencer from Boise would post pics with the brand's products in it . . .

And thus, the influencer marketing industry as we know it today was born. And for a while now, it's been a win–win for everyone involved. Social media influencers won because they found brands willing to pay big bucks for them to sit around and upload some pics to the 'Gram. And brands won because a social media influencer, despite their massive follower counts, is still lower down on the celebrity social hierarchy than A-listers, reality stars, models and musicians, and thus they demand less compensation than the bigger fish – yet the brand still gets the same audience reach it would have with the traditional celeb. By 2019, brands spent a staggering $8 billion annually on influencer marketing campaigns.[82] By 2022, that figure is expected to almost double to $15 billion *per year*.[83] However, where celebrity spokespersons once found themselves being pushed out of frame by 'ordinary' influencers, in recent years, a new breed of influencers have arisen that may see those human-based 'ordinary' ones shown the door, too.

This new breed is the virtual influencer, like Aliona Pole.

Right now, there are about 150 big-name virtual influencers, but their ranks are growing all the time to meet the demand from brands. While virtual influencers were once viewed as a niche oddity, in recent years, as their creators have carefully cultivated their social media presence, brands have seen the benefits of not just a virtual influencer's large follower count but the qualities inherently unique to their existence as a digital being – some of which seem almost godlike.

You see, unlike human social media influencers, a virtual influencer will never become less appealing to the next younger generation by aging (unless they want to); they will never embarrass the brand by doing something unacceptable in the real world, such as getting caught using hardcore drugs or becoming violent; they will never innocently or maliciously say the wrong thing in public, thus inciting the internet outrage mob; they will never get sick or say they can't do a sponsored post because they're on vacation or it's a holiday; they can speak and promote a brand's product in any language no matter how difficult it would be for a human to learn; and they can be not only in two places at once, but thousands of places – even in real-time.

Virtual influencers are also highly influential with the coveted 13-to-24-year-old crowd. Today, this group is a segment of the population who grew up intimately familiar with digital humans in the form of video game characters. For many young people today, following someone on social media who looks computer-generated is no odder than following a human social media influencer they've never personally interacted with before either. And though these virtual influencers are typically built using traditional CGI techniques Hollywood studios and video game developers have relied on for decades, their presence and actions on social media make them *feel* like real, relatable people even though they may not totally look it. This is because virtual influencers tend to post about the same stuff flesh-and-blood young people do.

For example, one popular virtual influencer is named Imma Gram.[84] Her bright pink hair is iconic to her over 330,000 Instagram followers who, judging by many of their comments on her posts, are quite invested in her 'life'. Though Imma's pink hair and perfect complexion suggests she's not like most of us, her Instagram posts beg to differ. Imma is Japanese and 'lives' in Tokyo. She posts about all the things any young person who lives in that city does: fighting with her sibling (also a virtual influencer), catching up on her reading, going to the newest trendy boutique in Harajuku for some shopping and, of course, taking pictures of her food at the latest cool café she's discovered.

Yet despite none of these things having actually happened in reality, her followers are quite obviously invested in her synthetic life nonetheless. In one Instagram post from January 2021, Imma is wrapped in a blanket, and her eyes are bloodshot as if she's been crying. The accompanying caption reads: 'It's been a whole month since my fight with my brother. I didn't know it was possible to go without talking to each other this long . . . Everything's so complicated, and I'm thinking about it way too much. How does everyone make up with their siblings?'

Despite Imma not having an actual brother, not having any sentience or awareness of herself, and not actually thinking about this fabricated scenario at all, there are over three-hundred replies to her post, with many followers offering their sympathies or advice.

One suggests that gifts usually work, so Imma should give her brother some food he likes or a present he's been wanting, and then apologise to him for the fight.

Another says Imma should tell her brother how she is feeling right now. If her brother doesn't care about her feelings or is unsympathetic, that's on him. There's nothing else Imma can do if she honestly tries to make amends but her brother won't listen.

A third doesn't offer Imma specific advice, but instead commiserates with her, explaining how they too hurt someone they cared for dearly and now don't talk with them, which leaves the commenter feeling empty and depressed – so they know what Imma is going through.

The responses like these – and many others – exemplify the deep connections young people have to these virtual influencers. This despite – based on a read of hundreds of comments across dozens of posts – a large majority of their followers knowing they're not real. And it's these deep *human* connections that make these virtual influencers so valuable to marketers. Brands love capturing the imaginations of these young teens and young adults, especially, because, though their incomes are currently low-to-non-existent, these young people will eventually grow into older adults with high disposable incomes – adults who are willing to separate themselves from their money for a nostalgic hit with a trusted name and face (just as we all do with Disney brands).

Marketers clearly see that value in Imma Gram. Like Aliona, Imma has had sponsorship deals with dozens of brands. Imma's roster includes the likes of Amazon, Nike, IKEA, KFC, Valentino, Calvin Klein and Dior. Or, rather, to be more accurate, the Tokyo-based 'virtual human company' Aww Inc. has those sponsorship deals. They're the company that created, owns, writes and directs the virtual life that Imma's hundreds of thousands of followers love to keep abreast of.

Though Aww wouldn't reveal how much a sponsorship deal with Imma costs, another company that owns a popular virtual influencer told me that a single sponsored photo post on Instagram could run a brand $6,000–$10,000 on average. A 30-second sponsored video on TikTok with that same virtual influencer, on the other hand, could cost a sponsoring brand as much as $50,000 to $100,000 for a *single post*, depending on what it entails.

It's that discrepancy between the cost of a photographic and video post that gives a big hint to one of the most significant drawbacks with virtual influencers compared to their human counterparts: the way virtual influencers are currently created is time-consuming and, thus, extremely costly.

Right now, your average virtual influencer like Imma Gram is created using traditional CGI techniques, which control their appearance, their expressions, the lighting that falls on them, and dozens of other details. All this traditional CGI trickery takes a virtual influencer's team of exceptionally skilled human artists hundreds of combined workhours hunched over their computers, painstakingly manually generating almost everything in the photo or video to the sponsoring brand's specifications. A single sponsored post with a static photo of a virtual influencer could take a team of digital artists as much as a week to create. A single sponsored post featuring a video of that virtual influencer? An even larger team of skilled artists taking three to four weeks.

In other words, though virtual influencers have significant advantages over their human counterparts, time is not one of them. While a virtual influencer's typical photo post takes a week, a human influencer can shoot and post a new photo in seconds. And forget the three-to-four-week turnaround for a virtual influencer's video post – a human influencer could get a thirty-second TikTok up in hardly longer than it took to record it – or at most a few hours if some editing is required.

Virtual influencers may be forever young, have perfect skin and speak dozens of languages but they can't react in near-real-time, getting content up at the speed of which some brands demand.

Or at least they couldn't.

And this is where we come back to Aliona Pole.

While it's true most of her Instagram photo posts are created using traditional CGI techniques, Aliona's Moscow-based creators use a more cutting-edge technology to hasten the time it takes for them to make a video post featuring the virtual influencer. It's a technology that cuts the production of a Tik-Tok video, for example, from weeks to hours – and thus gives Aliona a huge advantage over her virtual competition.

That technology is, of course, deepfakes.

Aliona is one of the few virtual influencers whose creative team has fully embraced deepfake technology. Called Malivar,[85] the virtual human company that invented Aliona is run by a team of just four co-founders. Early on, Malivar saw the potential in virtual humans, but they also immediately identified the tediously slow content creation process as a digital being's main problem – especially when creating content for video platforms like TikTok, which is where all the young and cool kids hang out today, and thus where all the brands want to be. And this slow process of content creation isn't just a problem for meeting the sometime-unrealistic deadlines sponsoring brands demand.

As any social media manager will tell you, if you want to get a large following on any online platform, the key is new content – consistently. It's not good enough to have one new post per week. You need one or more new posts *per day*, on multiple platforms, to keep followers engaged with your influencer's feed. It's that steady engagement that ultimately brings the sponsoring brands and their big bucks.

What Malivar's team realised in 2018 was that they might be able to get around the temporal creation bottleneck that plagues traditional CGI by adopting that new deepfake technology that had been getting all the bad press at the time. Forget porn, deepfake tech could be used to rapidly create virtual influencer content. But open-source software like DeepFaceLab didn't have the speed or scalability Malivar required for deepfaked virtual influencers. If they were to go down that route, Malivar would need to code their own deepfaking tool.

So that's what they did – and they called it Malivar.io. The bespoke software is a much more advanced version of the GAN technology that you find in DeepFaceLab. Its proprietary neural network and machine learning algorithms are far more intelligent, more automated and much faster than most other deepfaking software. As a matter of fact, Malivar's software is far enough removed from what you'd find in DeepFaceLab that the company doesn't even like to refer to what they are doing as 'deepfaking', instead referring to their software creations as 'neural masks'.

The distinction here is slight but important. Whereas Deep-FaceLab requires the user to manually compile facesets of actual human beings, which DeepFaceLab then creates a synthetic model of that it can map onto a donor body's face, Malivar's software allows facesets of already synthetic faces – like Aliona's CGI face – to be compiled and trained on. The result is a synthetic 'neural mask' of an already synthetic face. In Aliona's case, for her social media videos, Malivar records a human actress, say, working out on a yoga mat, and then slips a neural mask of Aliona on over the human stand-in's face.

Malivar's deepfake technology massively streamlines the content creation process of virtual humans. They can create multiple photo posts featuring the virtual influencer in a single day – or even multiple *videos* in the same day for platforms like TikTok. More posts mean more engagement, which means more brand sponsorships, whose own sponsored posts take much less time to create, thus generating more engagement and more money.

This is deepfaking for legitimate commercial profit.

The company, however, doesn't intend on solely using their Malivar.io-deepfaked Aliona to generate profits through sponsorships. Malivar realises that as the commercial uses of deepfaking become more prolific in general, and in the virtual influencer industry specifically, brands will increasingly want the tools to quickly create their own virtual influencers they can 100% own and control – and use to rapidly churn out a stream of engaging social media content. That's why the company has made its Malivar.io tools available to other businesses. Since Malivar.io's machine learning algorithms are so much more advanced than those of DeepFaceLab, it doesn't even require a human operator telling it what needs to be done. Instead, a brand can simply upload the video of the human actor they recorded and Malivar's software will automatically apply the neural mask of the brand's existing CGI influencer to the person in it.

Deepfake's original sin was that the technology was primarily used to create non-consensual pornography. As Malivar's shown, innovative companies are eager to apply the technology to generate synthetic content in a fraction of the time it once took – content that can help take a big slice of that annual $15 billion influencer pie brands are set to splash out on. In the process, applying deepfake technology in this way can help reform the public image of deepfakes, thus helping give the technology the legitimacy many in the influencer, video game, marketing, communications, and even healthcare industries feel its potential deserves. Yes, that's right. It's not just the virtual influencer industry that's set to reap possible billions in profit from deepfake-generated content. Deepfake technology is primed to add value to numerous other wildly disparate industries.

Already video game makers are exploring how deepfake technology can be applied to make photorealistic avatars of the individual gamer and insert their likeness into the world of the game itself. These deepfaked avatars need not be limited to the latest *Grand Theft Auto* though, they could also be used to give a body to the ethereal virtual digital assistants that live on our smartphones, smart TVs and smart speakers.

Deepfaked avatars could give Siri and Alexa corporal form – and not just one, but an infinite number of them, tailored to each user's preferred look and tastes. Imagine coming home from work and seeing Siri, perhaps with a red mohawk, or Vulcan-like ears, or even with your mother's face – whatever you like – pop up on your Apple TV to give you a visual update on your day, the house or global affairs. Walk into your kitchen and this visual version of Siri appears on the screen on your fridge door and talks you through making that egg soufflé you've always wanted to try.

Far-fetched? Hardly.

One company, Pinscreen,[86] enables you to place photorealistic synthetically generated faces over your own in real-time. Imagine never having to worry about looking good for a Zoom call again when you can simply slip on a ready-made version of your face, looking the best it's ever been.

In the fashion world, several companies are working on deepfake technology to show you exactly what a particular item of clothing will look like on you before you buy, by creating a deepfaked avatar of yourself to wear that sweater or jacket you've been eying – all from the comfort of your own computer. Say goodbye to returning your online purchases because they don't sit on your body the way they do on the website's model. And from the point of view of the brand, fewer returns are always a bonus and could save millions annually in return shipping costs, not to mention maintaining a much-happier customer base.

Now as someone on the younger side who is steeped in the tech world and enjoys near-future science fiction in the entertainment world, what I've described already in this chapter, to me personally, and probably to many in the generations younger than mine, sounds pretty cool. When I bring up this kind of deepfake avatar technology with people in generations older than mine, however, many have the same response: it unnerves them, the possibility that we'll be interacting with digital constructs in the way we do real people right now.

Many say they fear deepfake technology implemented in such ways will lead us to ceasing to interact with each other in the real world. That we'll turn to deepfaked avatars when we're lonely, instead of to one another. No one can predict if that will eventually happen, but I will point out that that's a common fear when any new interactive technology arises. People said that about the phone, then the internet, then social media – all these technologies would alienate us by stopping us from interacting in person or with others altogether. And to a degree, I think that has happened (especially in regard to social media more so than the others), though not at the level people feared. But I don't believe that we need to worry society at large is going to stop interacting when deepfake beings become commonplace. For one thing, while deepfake avatars may look hyper-realistic, the artificial intelligence proficiency they'd need to possess to replicate the interaction a human can provide us just isn't there. Yet.

I will also point out that the good deepfake technology can do isn't only limited to commercial enterprises. It's not just the gaming, communications, advertising and fashion industries that will benefit from deepfakes; it's not all about making money for big businesses. At the risk of sounding like a cliché marketing pitch, deepfakes can be used for the benefit of humankind too.

Synthesia,[87] for example, is a company that utilises GAN and neural network technology to allow businesses to completely synthesise entire people out of thin air with all the quality and verisimilitude of a video that looks like it was shot using real people with a real camera crew. All you do is plug in your script and your synthetic actor will say exactly what you want it to say – and not just in English, either. The synthetic presenter can speak in any of forty languages, meaning one video and one synthetically generated presenter could be used in markets worldwide. Synthesia's deepfake technology has already been used to enable the real David Beckham to speak nine different languages in a public service announcement to raise awareness of malaria in disparate regions around the world.[88]

Imagine that: deepfaking for improved public health. Deepfaking never seemed so good.

If we look further at the implications for the health field, you need only use a little imagination to see potential. Deepfake technology could radically improve the quality of life for some patients with debilitating conditions. As a matter of fact, it already is.

Project Revoice[89] is a non-profit initiative that uses deepfake audio technology to literally allow people to speak again. The voice cloning project is run by the ALS Association. People with ALS progressively lose the use of their body as time goes on, including the ability to speak. What Project Revoice does is use deepfake tech to clone an ALS patient's voice *before* they lose it and upload a copy of that cloned voice to their

Augmented/Alternative Communication (AAC) device. The ALS patient can then use their AAC device to continue speaking in their own unique voice.

It's important to understand that because of the versatility of the machine learning that underpins deepfake technology, these cloned voices are not simply a collection of pre-recorded phrases. They are the algorithmic digital essence of an ALS patient's biological vocal patterns, which the AAC device can emulate to articulate whatever the patient wishes. Imagine the suffering this specific application of deepfake technology relieves. It literally gives someone not just *a* voice back – but their *own* voice.

And the healthcare uses of deepfakes don't stop there. Think of the myriad reasons why those suffering from psychological trauma might be apprehensive about seeking help: social stigma; anxiety; concerns over anonymity. Simply sitting down face-to-face with a clinical expert can feel like an insurmountable challenge, but deepfake technology could allow a person living with PTSD to 'wear' a visually distinctive deepfaked face over their own during virtual counselling sessions, enabling them to seek the help they deserve and work through their trauma while mitigating their fears and concerns. And remember, since deepfaked faces mimic the expressions of the underlying real face, a therapist will still be able to pick up on the patient's non-verbal responses and nuances.

Deepfakes could also help people suffering from eating disorders. Using custom deepfakes personalised to a patient's look, someone with bulimia could see what they would look like with a slightly altered, healthier body. Allowing a bulimic person to slowly get used to looking at themselves in a healthier body could help change their distorted view of how a body is 'supposed' to look.

It's worth noting that deepfaking has one more obvious use that hovers between commercial and personal benefits: porn – and I'm not talking about the non-consensual deepfake celebrity or revenge porn that we've already discussed earlier in this book. Like it or not, porn has been around for as long as humans have been able to record themselves – from cave paintings through to the high-tech, high-production movies widely available today. Imagine if porn performers themselves had the option to use deepfake technology to overlay a synthetically created photorealistic face on top of their own. This would allow them to perform in the career they've chosen while enabling them to hide their true likeness from intolerant friends, family or others who would be quick to pass judgement on their life-choices. It would also allow them to not have their past follow them around if they chose to move on to other careers later in their life since their true face would have never been revealed. Deepfake tech could even allow porn performers new ways to create types of niche porn – such as cosplay. Does a porn star's followers like to see them roleplaying as an elf? Now porn performers can bring those fantasies to life for their fanbase without gluing fake ears to their head. In other words, deepfake technology could be used to make the world of porn more fair, more humane, more private and even more creative.

It's no shock that legislators want to legislate but given the wide-ranging uses for deepfakes – all the way from healthcare to entertainment – it's also not too surprising that any business that could benefit from the technology would have something to say about laws pertaining to deepfakes. So it's also not surprising that Disney would have a stance on that New York State bill introduced in 2017. Titled 'An act to amend the civil rights

168

law, in relation to the right of privacy and the right of publicity; and to amend the civil practice law and rules, in relation to the timeliness of commencement of an action for violation of the right of publicity' – but, mercifully, better known by its designating number A08155[90] – the bill was introduced in the New York State Assembly in 2017, undergoing several revisions before being put to a vote in June 2018.

As you can tell from its exceedingly long title, the main purpose of A08155 was to give a person new rights of privacy and rights of publicity over their persona – that is, their likeness, whether in visual or audio form. The right of publicity is something long enshrined in US law and, as the bill's justification itself points out, 'refers to every individual's inherent right to control the commercial use of his or her personal characteristics, which can include name, portrait or picture, voice or signature, each a part of an individual's persona.'

What A08155 specifically aimed to do was to create a right of publicity for deceased individuals that would extend for forty years beyond their death, while expanding current right of publicity laws to allow for anyone to treat their persona as a transferable or descendible asset – much like a house or a car. This meant under A08155, a person, be it a celebrity like Tom Cruise or an ordinary person like me or you, could sell or transfer the rights to their persona to another owner (be it a person or company) while they were still living, or upon their death. Essentially, you'd be able to asset-ise your own persona and, if somebody wanted to use any aspect of it, you could monetise that or leave it to your descendants to monetise.

So what does all this have to do with deepfakes?

Well, the timeframe during which A08155 was proposed (2017) and throughout the amendment process (into 2018) coincided with many of those alarmist headlines about deepfakes, such as *Vice*'s 'AI-Assisted Fake Porn Is Here and We're All Fucked'.[91] So what the bill's well-intentioned sponsors did was add a clause to A08155 in hopes of combating this emerging threat by forbidding the use of 'digital replicas' of a person without their or their heir's consent. This 'digital replicas' clause ensured that no one could argue the bill's provisions applied to only genuine images or videos of a person. The bill defined a digital replica as 'a computer-generated or electronic reproduction of a living or deceased individual's likeness or voice that realistically depicts the likeness or voice of the individual being portrayed'.

If there's a more succinct definition of 'deepfake', no one's found it yet.

And the bill didn't stop with a clause that simply ensured deepfakes would fall under it. A08155 went on to define specific scenarios that 'digital replicas' of other people were forbidden from being used in without that person's consent. Specifically, the bill stated, 'Use of a digital replica of an individual shall constitute a violation if done without the consent of the individual if the use is in an audiovisual pornographic work in a manner that is intended to create and that does create the impression that the individual represented by the digital replica is performing.'

What this meant, ladies and gentlemen, was that A08155 was the world's first law forbidding the pornographic deepfaking of another person – celebrity or not – without their consent. It would have made not only non-consensual pornographic deepfake videos illegal, but non-consensual pornographic deepfake photos as well – along with non-consensual fake pornography created using traditional digital tools such as Photoshop. Any production of non-consensual 'digital replica' pornography would be a clear violation of the bill and thus the deepfakers who created the non-consensual pornographic works would be liable for monetary compensation to their victims. Additionally, the bill didn't limit a victim's rights to pursuing additional legal avenues, whether criminal or civil, against the deepfaker.

In other words, A08155 was the law any victim of non-consensual deepfake pornography dreams of. And it sounds like a perfectly reasonable piece of legislation to propose and pass, doesn't it?

Yet ten days before A08155's final vote on the New York State Assembly floor, The Walt Disney Company came out opposing not just the bill but its non-consensual pornographic deepfake provisions, specifically.

Why? To reiterate, this is where Aliona Pole comes in. Not the virtual influencer herself, specifically, but what her creation, existence, virtues and use portend to the broader field of entertainment in the decades to come.

Though Hollywood studios have long used traditional CGI techniques to resurrect actors from the dead (as Disney did with Peter Cushing in *Rogue One: A Star Wars Story* and Universal did with Paul Walker in *Furious 7*), insert younger replicas of an actor into a movie (as Disney did with Carrie Fisher's nineteen-year-old likeness in *Rogue One: A Star Wars Story*) or de-age the actual actor onset (as Disney did with the then-seventy-year-old Samuel L. Jackson in *Captain Marvel*), those traditional CGI techniques are tediously time-consuming and extremely costly. And even then, audiences still often feel the resurrected or de-aged actor looks somehow inauthentic. We can tell who the CGI character is *supposed* to be, yet for some reason, it just doesn't *feel* realistic.

But with deepfake technology, it *feels* so much more real because of the way a donor body's real expressions are mathematically blended with a target's synthetically generated face. Remember the YouTube fancasting deepfaker Derpfakes's deepfake of Carrie Fisher from the first chapter? The one he created in less than thirty minutes? Disney's CGI'd version of Fisher's faces looked uncanny and lifeless. Derpfakes's deepfaked version of Fisher's face in that same scene looked natural and alive. And ironically, that natural look is thanks to *artificial* intelligence.

So, when it became known that deepfake technology was being successfully applied to make it realistically appear as if some of the biggest celebrities in the world were taking part in pornographic videos, some in Hollywood were already making the connection to what they could use the same technology for in movies. And that is making realistic digital replicas of their actors and actresses to use in ways even the most prolific porn deepfakers had never considered, but which the use of virtual influencers like Aliona Pole have already hinted at.

Such as? How about keeping an actor alive and acting in movies forever? If Aliona Pole can't die, why should Brad Pitt?

With deepfake technology, studios now know they never have to worry about scrambling to figure out what to do when an actor critical to a popular franchise dies – as what tragically happened with Carrie Fisher, Paul Walker and, most recently, Chadwick Boseman, who was scheduled to become the new face of the Marvel Cinematic Universe, and whose 2018 film *Black Panther* became the highest-grossing solo superhero film of all time.

Admittedly, it seems insensitive to be concerned with the financial ramifications to a studio when a beloved actor passes away, but as commercial enterprises, it's something Hollywood has to contend with – and quickly. This is why they've opened their pocketbooks and spent millions on traditional CGI techniques in the past when an actor who that story relies on unexpectedly passes away. It's an attempt to salvage films they've already spent hundreds of millions on, which could end up being worth billions. Deepfake technology could eliminate the risk of an actor's untimely demise entirely for a studio. And when billions of dollars are at stake on not just one film but its inevitable sequels, you can bet a technology that easily allows studios to resurrect the dead is something they will not discard.

But why stop at resurrecting dead actors? Just as deepfake technology allows Aliona Pole to be doing different things in multiple places at once, the technology could easily allow for a single actor to do the same. Leonardo DiCaprio is one of the most in-demand actors in the world. If he wants to make a movie, any studio will greenlight it because DiCaprio puts butts in seats. Unfortunately, there are physical limits to how many DiCaprio films can be made. After all, there's only one him, and he can't be shooting a Western in Australia and a sci-fi thriller on a lot in Burbank at the same time. Or, at least, he never used to be able to. With deepfake technology, a studio could shoot an entire film with a stand-in for DiCaprio and then simply deepfake DiCaprio's actual face onto the face of the 'donor' actor. Hell, why stop at two movies at once? Why not four DiCaprio films at once? Ten even? The more there are, the more the studio makes.

Another benefit of Aliona Pole? She never ages. Well, unless she wants to. She could slip on one neural mask and be 25 today, another and be 47 on Tuesday and yet another and be back to being 18 in time for the weekend. So why shouldn't human actors be able to do the same?

Does a studio want a romcom with a 20-year-old Keanu Reeves? Now it's doable. The 57-year-old Reeves could act in it and deepfake tech could plop a neural mask of his 20-year-old face on top of his real one in a fraction of the time and much more realistically than it would take using traditional CGI technology. In this way, deepfake technology actually eliminates 'age-appropriate' roles for actors. The acting industry is one where ageism is rife at times – that is, many actors are turned down for roles because they are perceived as being too old. But with deepfake tech, a 65-year-old actress could play a 25-year-old college student and bring to the role all the skills, finesse and experience the actress's real 25-year-old self could have never had at that stage in her career.

The possibilities of how deepfakes can be used don't just end there. Just as Aliona Pole can speak forty languages if needed, so too now can Hollywood actors. Using deepfake audio cloning tools, studios could do away with dubbing and subtitles entirely. Instead, they could simply deepfake the actors on the screen to make them speak French or Japanese or German or any other language they wish. This would vastly increase the appeal of Hollywood movies in increasingly important foreign markets – especially those in China. If Gal Gadot could suddenly speak flawless Mandarin, Chinese audiences might be much more likely to turn out in droves for the latest *Wonder Woman* film. Such voice cloning could also encourage filmgoers in one area of the world to experience more films from other cultures. I've known people who have passed up seeing classics like *Crouching Tiger, Hidden Dragon* and *Rashomon* because the films are only offered subtitled from their original Mandarin and Japanese, respectively. Imagine the cultural awakening and understanding that foreign films could rouse in overseas audiences if the actors in them spoke the local language flawlessly.

Of course, despite the huge commercial advantages deepfake technology offers for Hollywood studios, it's worth asking what the application of deepfake tech means for the next generation of actors. If Will Smith and Jennifer Lawrence can now live forever, speak any language and look any age on demand, will the industry ever need new superstars? Or perhaps future 'actors' will simply be relegated to being puppeteer donor bodies, renowned for their physical movements and facial expressions, not their identity. Such a transformative change could birth a new Academy Award category: Best Deepfake Puppeteer.

Given all the opportunities deepfake technology opens for Hollywood, it's hard not to get excited about how the tech could usher in the next era of filmmaking. Deepfake technology could be as transformative to the film industry in the 2020s as sound was in the 1920s, colour cinematography was in the 1950s and computer graphics were in the 1990s.

It's now easy to see why The Walt Disney Company has such a keen interest in any laws or regulations surrounding the use of deepfake technology. As a matter of fact, Disney hasn't just shown that it's concerned about laws that could limit the use of deepfakes, the company is actively working to improve the technology itself.

Right now most deepfakes have one major technical limitation: the deepfaked faces in the videos can't be any larger than 256 pixels high by 256 pixels wide. This resolution is adequate quality for the deepfaked faces appearing in the fancasts posted to YouTube and in fake celebrity porn because those videos are typically viewed on small screens — smartphones and laptops. But if you projected these deepfakes onto a cinema-sized screen, the deepfaked faces would look blurry. They'd be too lo-res compared to the sharpness of the rest of the massive picture.

So what's Hollywood to do if they want to eventually be able to make billion-dollar-blockbusters with the world's hottest actors for the rest of time? Well, they need to work on improving the quality and size of deepfakes themselves. And that's just what computer scientists from Disney Research Studios (yep, that's a real thing) have been working on.

In 2020, they announced[92] that using new neural algorithms and other advanced machine learning techniques they were able to increase a deepfaked face's resolution *sixteen*-fold. As the company's researchers showed, Disney's deepfakes look sharper, more detailed and have no discernible seams or artefacts. Of course, Disney's researchers aren't developing their in-house deepfake tech to help seedy deepfakers create higher-quality fake celebrity porn. The endgame for Disney is to help the studio make better films with more realistic synthetic imagery in much less time than traditional, costly, CGI techniques take.

Now, Disney's researchers are still experimenting with their deepfake technology improvements, and they still have some other bottlenecks to overcome before Disney can start pumping out a new trilogy of *Star Wars* films starring a young Mark Hamill back in the role of Luke Skywalker. But as Disney's researchers have shown, getting there is no longer an impossibility – it's an inevitability. And fans of the original *Star Wars* trilogy will be happy to hear that at the rate Disney and Hollywood in general are experimenting with and iterating upon deepfake technology, within the decade deepfake tech could be as commonplace in Hollywood blockbusters as CGI is today. With that will come a slew of new films featuring beloved characters we never want to see retired.

Given Disney's work in the deepfake field, it's understandable that the company has a vested interest in making their voice heard when any legislation is introduced that could potentially regulate or place limits on the technology's use. Still, New York State Assembly bill A08155 wasn't proposing to limit deepfake technology usage in general. Rather it specifically wanted to make the creation of non-consensual deepfake pornography illegal. So why did Disney still come out in opposition to that? After all, the House of Mouse doesn't make skin flicks.

The answer is revealed in a letter[93] Lisa Pitney, vice-president of government relations at The Walt Disney Company, sent to New York State lawmakers on 8 June 2018 – ten days before those lawmakers were to put the final version of the bill to a vote. The majority of that letter is focused on Pitney's 'serious concern' that the 'broad and new right of publicity' changes the bill proposed could have on film, television and other creative productions in the state. Pitney does go on, however, to address A08155's intention to place a legal restriction on the creation of sexually explicit deepfakes:

> On top of all this, the bill would create entirely unprecedented rights to control the use of 'digital replicas' and the use of celebrity images in sexually explicit material, which while presumably well-intended threaten expressive activities as a result of undefined, vague or otherwise problematic statutory language.

In other words, Disney seemed to feel the statutory language in the bill could not only prevent shady internet deepfakers from creating non-consensual pornographic videos, but it would inadvertently make it illegal for companies like Disney and other film studios to feature a deepfaked representation of a deceased actor in not just a sex scene, but scenes featuring nudity or of a sexual nature. As Pitney's letter alludes with its reference to 'undefined, vague or otherwise problematic statutory language', A08155 stated it would be illegal to feature an individual's digital replica in a 'pornographic work' without that individual's consent. But what is a 'pornographic work'? Does it mean a filmed sex scene that explicitly shows penetration? Or is a nude scene that just implies the act of penetration enough to be pornographic? What about a scene with no sex but just frontal nudity? If an actress gets out of the shower and you see her breasts, is that pornographic? Is fully clothed groping pornographic? Is the scene of two men French-kissing? Is the scene of a mother breastfeeding?

Depending on who you are, where you live, the culture you grew up in and how you were raised, you may see one or more of those filmed scenarios as pornographic. But 'pornographic work' actually has no fixed definition in law. Furthermore, the bill didn't distinguish between 'pornographic work' in an artistic piece, such as a film, and a 'pornographic work' that is made solely for one's sexual gratification, such as the typical pornographic celebrity deepfake.

If it seems like Disney is being nitpicky, it's not. Many laws have unintended consequences because of how their statutory language is worded or because their full implications haven't been thoroughly considered by lawmakers eager to tackle a burgeoning problem. Disney was just trying to stop an unintended consequence from affecting a filmmaker's or studio's ability to tell a story – especially biopics of deceased individuals – before it was too late. Once a law is passed, it could take years, if ever, for it to be amended to remove or clarify the language that introduced the unintended consequences – and in that time, Disney would have no choice but to change the stories they wished to tell if those stories relied on a scene that could be considered 'pornographic' with a deepfaked actor or real-life subject who had passed away and thus could not provide their consent.

'If adopted, this legislation would interfere with the right and ability of companies like ours to tell stories about real people and events,' Pitney's letter to the New York State lawmakers went on. 'The public has an interest in those stories, and the First Amendment protects those who tell them.' The letter concluded: 'For these reasons we ask that you oppose this bill as drafted and vote no should it come before you.'

It's worth noting that Disney isn't the only major player that came out against A08155. Warner Bros. Entertainment, Viacom, Getty Images, the Entertainment Software Association and the Motion Picture Association of America, among others,[94] also did.

In the Motion Picture Association of America's (MPAA) letter,[95] the body noted, 'While MPAA agrees that the emerging "deepfake" problem deserves attention, the hastily drafted provision in the Bill would result in unintended, harmful consequences and is likely unconstitutional. The provision governs uses in "pornographic work[s]", but the Bill does not define that term, which is extremely vague and lacks any accepted legal meaning. Also, there are no exceptions or limitations in the Bill that would exempt uses in clearly First Amendment-protected contexts including news reporting, commentary and analysis.'

The MPAA's letter, along with letters from NBCUniversal[96] and the Electronic Frontier Foundation[97] opposing the legislation, were delivered to New York Assembly legislators on the same day as Disney's letter – 8 June 2018. Regardless, ten days later, on 18 June 2018, the New York State Assembly held a vote on A08155. The results weren't even close. A08155 overwhelmingly passed with 131 'yea' votes to nine 'nay' votes.

Yet this anti-non-consensual pornographic deepfake law was never to be – at least not in this iteration.

Though the New York State Assembly passed A08155, its companion bill, S5857B,[98] still needed to be taken up by the New York State Senate before the provisions could become law. Yet, the New York State Senate adjourned its legislative session two days later, on 20 June 2018, before taking up the companion bill. When the next legislative session began that autumn, the New York State Senate did not reintroduce S5857B again.

Yet all was not lost. In a film-worthy epilogue – in November 2020, New York's governor did sign another bill into law, the New York bill mentioned in the previous chapter, which arose out of the ashes of A08155. Designated Senate Bill S5959D,[99] that bill establishes the forty-year right of publicity after a person's death that A08155 had sought. More importantly though, S5959D also explicitly banned the non-consensual pornographic deepfaking of an individual. This law now allows civil action to be taken against deepfakers who use a person's face in 'sexually explicit material' without their *written* consent – which is more proof than the deepfaker would have required under A08155.

However, S5959D includes specific exceptions to this provision – which likely reassured all the Hollywood interests who fought against the passage of A08155. Those provisions say that sexually explicit deepfakes that are a work 'of legitimate public concern, a work of political or newsworthy value or similar work, or commentary, criticism or disclosure that is otherwise protected by the constitution of this state or the United States' are exempt from the law's ramifications provided 'that sexually explicit material shall not be considered of newsworthy value solely because the depicted individual is a public figure'.

What this means in practice for *celebrity* deepfake pornography remains to be seen – if any challenges by celebrities against a deepfaker go to court, First Amendment freedom of speech guarantees could still lead to the suits being tossed. But for 'ordinary' people out there who are deepfaked into sexually explicit material without their consent – S5959D is a major step in the right direction.

From virtual influencers to fashion to healthcare and most certainly to the entertainment industry, the commercial potential of deepfakes goes a long way in explaining why companies such as Disney would be apprehensive of laws placing restrictions on the technology. Yet if companies are on the cusp of doing so much with deepfakes, it makes you wonder what even more powerful institutions could do. Specifically, what could a nation-state wielding deepfake technology achieve? For that we need not look to the future, but into the past . . .

Chapter Six

The End of History

When you're a kid growing up, you're taught that history is fact. That is, the history you read about in your textbook is one hundred per cent how things played out. It's only when you're older and perhaps better travelled, or at least more widely read, that you realise just how subjective history is. Or, to put it another way, you realise how true the phrase is 'history is written by the winners'.

But with deepfakes, even that's no longer the case. Now the records that underlie modern history can be written – or more precisely, rewritten – by virtually anyone; and rewritten authentically enough where they could pass as historical fact. To give you an idea about what I mean, I want to tell you about the best deepfake I've ever seen.

It's of the moon landing in 1969.

We all know the real story – on 20 July 1969, the Apollo 11 Lunar Module *Eagle* touched down on the moon. It contained the first two humans ever to reach the lunar surface, Neil Armstrong and Edwin 'Buzz' Aldrin. It was truly a historic day in human history. For the political and military leaders of the United States of America, though, this day meant more than just a milestone in human achievement. You see, as far as they were concerned, the most important thing about this event was that it was an *American* achievement. *Americans* Armstrong and Aldrin were the first people ever to reach the moon, and they got there via an *American* spacecraft and an *American* space programme. And in doing this, the Americans had beat their economic, political and social arch-rival the Soviet Union in the greatest race in history – the Space Race. In the midst of the Cold War, this was the most consummate victory the United States could dream of short of defeating the Soviets in actual armed conflict. Up on the moon, six-and-a-half hours after the Lunar Module *Eagle* touched down, Armstrong planted his foot onto the lunar surface and spoke one of the most iconic phrases of all time: 'That's one small step for [a] man, one giant leap for mankind.'

The American's words were heard by over 600 million people back on earth.

That's how things actually happened. Yet the best deepfake I've ever seen shows how things might have turned out. Titled *In Event of Moon Disaster*, the film shows how, perhaps, the Soviet leadership of the day wished things would have played out on 20 July 1969. As the seven-minute film begins, we see CBS News coverage of the Apollo 11 mission. We see news legend Walter Cronkite narrate us through Apollo 11's lift-off and the three-person crew's journey to the moon. We see footage of the Apollo astronauts inside the command module, and we see the lunar module containing Armstrong and Aldrin descend to the moon's surface.

But suddenly, the footage jerks.

And we see the lunar surface outside the lander's window.

The landscape is slanted, a diagonal with grey rock on one side and black space on another.

Something's gone horribly wrong.

A staticky voice from Mission Control says, 'Lost data flight.' Another person replies, 'We've had shut down.'

Without warning, the broadcast we're watching cuts to a television test pattern. You know the one: the rainbow colour bars that any terrestrial television signal in the mid-twentieth century would cut to when a station abruptly severed its feed.

It's a moment before the colour bars give way to a CBS News SPECIAL REPORT logo onscreen. Then there's a shot of the exterior of the White House. It's dark outside. A lonely American flag sways from a pole in the night.

Cut to inside the White House. We see an empty chair and desk in front of a blue backdrop. On the desk are two microphones. On either side of the chair are flags; one the red, white, and blue; the other the flag of the President of the United States. There's indistinct chatter as a few personnel adjust the objects in the frame. Then there's a cut. We're on a close-up of the chair now. A man comes into frame holding a few white sheets of paper. He sits down. It's President Richard Nixon.

He looks sombre. Even a bit ashen, though that could be because of the historical nature of the footage. It's over fifty years old, and back then, television footage wasn't as sharp and crisp as it is now due to the analogue techniques used at the time.

'Good evening, my fellow Americans,' Nixon says as he looks from the camera back down to the sheets in his hands. His voice is probably the most iconic of any political leader. You just have to hear a few seconds of that deep throaty tone, and you instantly know who it is. For a moment, Nixon says nothing more, simply gazing at the white sheets in his hands. It's as if he can't believe what he's about to read. Finally, he continues:

'Fate has ordained that the men who went to the moon to explore in peace will stay on the moon to rest in peace.'

Another pause to collect his thoughts.

'These brave men,' he goes on, 'Neil Armstrong and Edwin Aldrin, know that there is no hope for their recovery. But they also know that there is hope for mankind in their sacrifice.'

And that's when you feel the hit in your gut. Now you understand what's happened – what the footage you just saw actually showed. The Lunar Module crashed onto the moon. That's why the lunar surface was tilted through its window. Armstrong and Aldrin survived, but they will slowly suffocate to death because the damage the lander sustained during its crash means it cannot lift off from the moon's surface. As Nixon is reading this, those two men are trapped inside a small tin can on a rock 238,855 miles from earth. There will be no rescue mission.

'These two men are laying down their lives in mankind's most noble goal: the search for truth and understanding. They will be mourned by their families and friends; they will be mourned by their nation; they will be mourned by the people of the world—'

You hear Nixon choking up. You can hear it in his throat and see it in the way his mouth frowns at the edges, the way he nods and pauses and nods again before continuing.

'They will be mourned by a Mother Earth that dared send two of her sons into the unknown. In their exploration, they stirred the people of the world to feel as one; in their sacrifice, they bind more tightly the brotherhood of man . . .'

Nixon pauses again. He looks into the camera the way one does when they need a moment, for if the next words are spoken too quickly, the world might see the President of the United States weep. As the camera slowly zooms in on Nixon's face, he looks back down to the sheets of paper held in his hands. His head shakes.

'In ancient days, men looked at stars and saw their heroes in the constellations. In modern times, we do much the same, but our heroes are epic men of flesh and blood.' Another pause and a tightening of his throat. 'Others will follow, and surely find their way home. Man's search will not be denied. But these men were the first, and they will remain the foremost in our hearts. For every human being who looks up at the moon in the nights to come will know that there is some corner of another world that is forever mankind.'

For a moment, Nixon looks directly into the camera again. It's as though he wants to add something off-script. Yet his face simply trembles, and he says, 'Goodnight.'

Cut to the seal of the President of the United States and then to black.

Now thankfully, of course, this televised address, despite looking like authentic historical footage from the CBS News archives, never happened in reality. That's obvious to me because I know how the actual moon landing turned out. Yet, if I had no knowledge of the moon landing, which happened over half-a-century ago now, and someone simply showed me that clip, I would say, without a doubt, that Nixon's televised address I just watched was 100% authentic archival footage of the US president's address to the nation on 20 July 1969, after America's disastrous attempt to land men on the moon.

That's how good this deepfake was. And as you know by now, I've seen a lot of deepfakes in my time. Yet this one was different. It was completely flawless to my eyes. There was no 'tell' – no moment where the face slipped for a few frames or where Nixon's blinks didn't match normal patterns. His voice, his expressions, his pauses, they looked completely authentic, indistinguishable from true historical fact.

Even knowing better, knowing everything that's possible – and even knowing those things that are only on the cusp of being possible – my brain was telling me that this was *real*. If you've never been duped in this way, it's hard to describe the unnerving sensation of this mismatch between what I was seeing with my own eyes, hearing with my ears, and what I knew to be true. The technical proficiency, the way Nixon's lips moved so naturally, and especially the way his voice – that instantly recognisable deep, throaty voice – emoted true despair. It was deepfaking at its best.

How was it possible?

In Event of Moon Disaster is an experiment created by two media artists, Francesca Panetta and Halsey Burgund.[100] Panetta is the former creative director at the Massachusetts Institute of Technology's Center for Virtuality and Burgund is a fellow at MIT's Open Documentary Lab. The duo originally came up with the idea in 2019 when brainstorming audiovisual projects related to the fiftieth anniversary of the moon landing. They were aware that years earlier, the US government released an alternate address from its archives, a speech penned in 1969 for Nixon to deliver should the moon landing end in catastrophe. That speech was called 'In Event of Moon Disaster' and was drafted for Nixon by his speechwriter William Safire.[101]

Thankfully, the real Nixon never had to deliver this alternate address – yet its existence meant Panetta and Burgund had the subject for their audiovisual project for the moon landing's anniversary. As artists, now that they had their subject, they then asked themselves the question all artists ask when they want to tell a captivating story: what's the best way we can bring the tale of this alternate speech to life so the audience engages with it on an emotional level?

That's when they settled on deepfakes. After all, what's more emotional than being able to present this tragic address as if it had actually happened? This, of course, was in 2019 during the height of the alarmist headlines circulating about the technology, the very real threat of non-consensual fake porn, the theories that deepfakes could be used to sway the upcoming 2020 US election. But as multimedia artists, Panetta and Burgund also recognised the technology's potential from a creativity and educational standpoint, and that's when inspiration really struck. Through the use of deepfake technology, Panetta and Burgund could, at the same time, bring a little-known part of American history to life (the alternate Nixon speech) while also educating people about emerging technologies that can alter and obfuscate the media all around us.

'We wanted to create what we call a "complete deepfake"; that is, one that synthetically produces not only the visuals, but also the audio,' Burgund tells me. 'We also wanted to use as much authentic material as possible, changing only the parts of the source material necessary to get Nixon to read the contingency speech. The more "real Nixon" we retained, the more convincing our "synthetic Nixon" would be.'

But how could the duo make a 'complete deepfake' that felt so real that anyone who didn't know the true history of the moon landing would think they were watching authentic historical footage of a sullen Nixon addressing the nation? DeepFaceLab? No. DeepFaceLab is great for swapping faces into movie clips and porn videos, but the software can't handle the complex and nuanced facial movements that a president delivering a televised address exudes. Besides, DeepFaceLab can't put words into people's mouths.

If they wanted their fabricated Nixon address to look and sound real, Panetta and Burgund knew they needed to turn to the professionals. And I'm not talking about deepfakers for hire you can find in internet forums. I'm talking about the computer scientists, engineers and creative professionals who loathe the term 'deepfake'. For them, it's 'synthetic media' – the technical term for the technology before porn got its grubby hands all over it. These are the types of professionals who work at startups in the burgeoning industry of synthetic media creation – the people commercialising deepfake technology for legitimate uses. And to truly bring *In Event of Moon Disaster* to life, Panetta and Burgund decided not to turn to just one of these startups, but two. One would handle the visuals, bringing a synthetic facsimile of Nixon to life, while the other would handle the audio, synthetically constructing that instantly recognisable throaty voice.

For the visuals, Panetta and Burgund approached Canny AI.[102] The Israel-based startup isn't a full-service 'deepfake' company per se. They won't put your face on Superman's body. Instead, they specialise in a specific type of synthetic media creation known as Video Dialogue Replacement (VDR). The 'dialogue replacement' part simply means Canny AI can take the recorded dialogue a client provides and insert it into the mouth of someone who didn't actually speak it. Yet it's the 'video' part of VDR where the synthetic media magic really happens.

Dubbing someone else's voice over a person in another video is something I can do on my laptop using built-in apps and my basic audio and video editing knowledge, but such a method is crude. Why? Because dubbing audio over someone else in a video using traditional techniques doesn't look realistic at all. The lip movements of the target's face don't match the words they've been forced to speak, nor the movements of the jaw and surrounding facial muscles. In other words, traditional dubbing techniques look completely fake. What Canny AI does is apply proprietary deep learning algorithms to a target video to learn and then synthetically reconstruct the target person's face. Then, when fed video of an actor speaking the desired audio to be inserted into the target person's mouth, those same algorithms automatically map the actor's facial movements onto the target's in a way that realistically mimics the natural movement of the target's lips, jaw and facial muscles as they now speak each word of the desired dialogue.

Normally, firms turn to Canny AI for commercial purposes. For example, an international agriculture company might shoot one marketing video but want to have the spokesperson in the video speak multiple languages so it can be distributed in as many global markets as possible. Canny AI creates the same video of the spokesperson speaking English, Spanish, Chinese, Taiwanese, Farsi – you name it. The synthetic manipulation of the spokesperson's face is so realistic you would swear they were a polyglot – their lips and surrounding facial structures move as they would have had they actually spoken the words. Canny AI are at the top of the synthetic media food chain – that's how good they are.

Panetta and Burgund decided to approach them to create their Nixon video to fabricate a moment in history that never happened. All they had to do was supply a target video – and as their subject was a former President of the United States, there was no shortage of footage to work with. The problem was the 'In Event of Moon Disaster' address was sombre in tone and, most of the time, Nixon's public addresses tended to be on the more aggressive or macho side.

'We spent many, many hours watching various Nixon speeches, most of which were of him posturing to the nation and the world about America's stance in Vietnam,' Burgund recalls. 'And then we saw his resignation speech from 1974, and we knew we had a winner. Not only did Nixon's entire demeanour feel suitably emotional for the contingency speech, but there was a wonderful zoom part way through the clip that revealed an up-close and unvarnished countenance.'

The next challenge was possibly more problematic: replicating one of the most distinct and recognisable voices in history.

If the subject of Panetta and Burgund's deepfake was an ordinary person, they could have had any voice actor record the unused address and VDR that, but *President Nixon* needed to be reading this address. And the goal of *In Event of Moon Disaster* was to be as realistic as possible while also demonstrating the complete creative and manipulative power of synthetic media – the way it can unreservedly fabricate video and audio. Because of the realism required and the desire to demonstrate the fullest capabilities of AI-enhanced synthetic media, Panetta and Burgund knew they didn't want to fall back on the professional impersonator, the standard Hollywood had used for decades.

'From a conceptual standpoint, the artwork would not be nearly as strong if we didn't go whole-hog with synthetic media,' Burgund says. Nixon impersonators were out, in other words. If this was going to be a 'complete deepfake' like they envisioned, Nixon's voice needed to be completely synthetic too. 'We wanted to put our money where our mouth was, so to speak.'

That's where the Ukraine-based Respeecher[103] came in. This 'voice cloning' company uses proprietary deep generative modelling techniques to transform one person's voice into another's – regardless of whether the two sound alike. Voice cloning is the type of deepfake audio technology that allowed scammers to make off with $240,000 of the British energy firm's money (Respeecher's technology had nothing to do with that audio deepfake heist). Respeecher's clients, on the other hand, include the likes of Lucasfilm and Digital Domain, one of Hollywood's most celebrated visual effects houses, which goes to show how sought after their deepfake audio expertise is. To get the most realistic clone of Nixon's voice, however, Respeecher would need to train its tech on not just Nixon but also the actor who Panetta and Burgund hired to read the original 'In Event of Moon Disaster' address. That actor was Boston-based Lewis D. Wheeler.

The fact that it didn't matter that Wheeler didn't sound a thing like Nixon highlights the power and potentials of deepfake audio. Respeecher's voice cloning AI would take care of transforming the actor's voice into the late president's. What was critical, however, was that Wheeler could fully slip into the persona of Nixon and embody the essence of the way he communicated. Or to put that another way, Burgund describes Wheeler's role not as an imitator, but as a deepfake 'puppeteer' and likens the process as something akin to the motion capture techniques used to bring CGI characters like Thanos in the Marvel movies to life, but for audio instead of visuals.

'Lewis really was able to "envoice" the character of Nixon and create a performance that had the appropriate sombreness as well as Presidential-ness. The speed, cadence, and, crucially, the pauses, all helped create a critical feeling of authenticity,' Burgund explains. 'And that is reflected in our final piece – but there's not an iota of his actual voice. The voice is completely synthesised.'

Wheeler spent days in a studio listening to Nixon recordings to learn the essence of how the president spoke and to record his own snippets of Nixon's audio – sometimes entire phrases, sometimes just a word – sometimes not even complete words to create a library of parallels; authentic Nixon audio and new Wheeler audio, that would ultimately be used by Respeecher to transmute anything Wheeler recorded into Nixon's voice – like the full address. Simultaneously, Canny AI used a video recording of Wheeler delivering the speech to align Nixon's resignation speech facial motion to that of Wheeler.

From there, Panetta and Burgund took all the various synthetically generated audio clips – and there were many – and painstakingly selected the best ones, both for audio quality as well as cadence fit with the Canny video, applied some manual edits and combined them. And so, the duo now had audio and visuals of the speech that never happened. Finally, Panetta and Burgund used archival footage to create the full narrative of the seven-minute film, applying some manual edits before editing them to form the final product. The result is *In Event of Moon Disaster*.

'As artists,' Burgund explains, 'our primary hope for the project has been to create an immersive and moving experience for people. But one of the great things about art is its power to deliver more than just the experience itself. We hope that putting our audience in a heightened emotional state – by stranding beloved astronauts on the moon! – they will be ready to take away lessons about how deepfakes are created and how powerful and manipulative they can be.'

Not only is Panetta and Burgund's deepfake the most convincing I've seen, it's also the noblest, a respite considering what the majority of deepfakes entail. The duo's deepfake is a kind of civic engagement project to both inform and educate others about the potentials of the technology – the good and the bad.

That being said, it is more than a little frightening just how authentic *In Event of Moon Disaster* looks and sounds. I know if I'd show it to my youngest nieces today, who aren't yet old enough to have learned about the Space Race or how the moon landing actually turned out, there's no reason I couldn't pass *In Event of Moon Disaster* off as the authentic record of a historical event. And that in itself is unnerving, this physical manifestation of altered history I can now play for anyone right from my smartphone.

If there's any solace, it's that Panetta and Burgund did not create *In Event of Moon Disaster* by themselves. They worked for months on the seven-minute film, with two separate companies – experts in synthetic media, both of which have strict policies against their services being used to create deepfakes for nefarious purposes. On top of that, the duo worked with a team of other multimedia professionals, including video and sound editors, re-recording mixers, and colourists to bring the film to life. That's to say nothing of the fact that Panetta and Burgund are multimedia maestros to begin with, with decades of experience between them, and they used their manual, human-generated talents to fine-tune the final product. The point is, for a deepfake this authentic and convincing, it takes a large, talented team of experts. Your average person sitting in a basement with DeepFaceLab on his computer isn't going to pull off a deepfake of this quality.

The worrying thing is there are other entities out there with easy access to teams of experts much larger than Panetta and Burgund's, and which have an almost unlimited amount of financial and technological resources at their disposal, allowing them to pull off deepfakes of the quality of *In Event of Moon Disaster*. I'm talking about nation-states.

Thanks to the financial, talent and technological resources nation-states possess, they truly have it in their power to wield deepfake technology in such a way that it could become a sort of de facto alternate history machine, cranking out fabricated historical records at will. And it doesn't take much imagination to think up dozens of potential scenarios that could be claimed, then backed up with video evidence as actual, historical truth. I'd love to be able to say that the nation-states of today would use their hefty power and resources for noble ends rather than creating alternate versions of past events. But we all know that's probably not true for many of them. After all, what does any government want to be able to do? Control the current narrative, of course. This technology gives them the tools to do that to a spectacular, terrifying degree.

One state could create footage of another sanctioning acts of terrorism that could lead to full-scale war.

A government could produce a video of their opposition party's leader, back in his younger college days twenty years prior, partaking in a march supporting radical and dangerous ideologies.

A populist president that demonises a certain group of 'undesirables' could share altered footage of previous, admired, presidents doing the same – thus arguing his claims are supported by historical fact: 'These people have always been loathed by our country and its greatest leaders!'

Manipulation of the past at this level admittedly sounds like something out of George Orwell's *Nineteen Eighty-Four*. Sadly, we need not turn to fiction for examples of the fabrication of historical records by those who control nation-states. Numerous bygone dictators have attempted to alter historical facts through media manipulation for their personal and ideological gain.

One stands out above all the rest.

Joseph Vissarionovich Stalin was arguably the worst dictator in history. Certainly, the worst Russia had ever seen. He ruled the Soviet Union from 1927 until 1953 with an iron fist. During his tenure, as many as 20 million Russians were murdered or executed[104] (though the exact number is still hotly debated). In one year alone, during 1937–8, as many as 1.2 million Russians were killed in what is now known as The Great Purge. It was called a 'purge' because many of the people killed were not just political enemies and so-called 'undesirables', but rather communist party members, Stalinist government officials and Red Army leadership – about whom Stalin had become paranoid. It was these high-ranking officials that, once 'disappeared', were never to be mentioned or referred to again – lest you wanted to disappear, too.

The problem for Stalin, unlike for history's previous dictators, was photographic media. It was fairly abundant by the 1920s thanks to newspapers, magazines and books, not that this in itself worried Stalin. As a matter of fact, like most dictators, Stalin wasn't only in love with himself, he was in love with his likeness and was happy to have it portrayed in the day's contemporary formats – plays, statues, paintings and, yes, photographs – as much as possible (provided it was always in a positive light). Stalin's problem with photographs only came when he started having his ex-comrades brutally murdered.

You see, many of the early photos of Stalin featured himself surrounded by his political and military confidants – confidants, there was a high chance, who would end up 'disappeared' one day by order of their friend and leader. When that happens, the last thing a dictator wants is for the people to come across an old photo and recall the ex-buddy the dictator's bumped off. If they did, the people might not just remember the disappeared person, they might begin to wonder what kind of monster could kill his own comrades. They might make those disappeared comrades into martyrs and conclude that the smiling man in the photo who ordered the death of his friends was a devil to be dealt with.

Before the widespread use of photography in publications in the early twentieth century, this wouldn't have been a problem. Just not many people back then saw photographs regularly. And without photos of this Communist party member or that Red Army colonel general, those who didn't know them personally would soon forget their faces and eventually names – if they ever knew them at all. But the abundance of printed photography from the 1920s onwards meant a disappeared person could live on forever in the eyes of everyone, and their likeness could grow in meaning and importance in the survivor's minds. And as time went on, and more of his once-close comrades 'disappeared', this possibility became a significant worry for Stalin, especially as by the late 1930s, there were official photographs in existence featuring the smiling dictator surrounded by five or more of his closest confidants, most of whom he had had killed by then. So, what was Stalin to do?

Launch the most extensive photomanipulation operation of the twentieth century. And the aim of that photomanipulation operation was historical negationism, that is, a distortion of the historical record – an attempt to fabricate the past.

For Stalin, this meant removing all his dead buddies from official government photos – a seemingly monumental task. Yet showing just how much fear Stalin commanded, the dictator didn't need anything close to *Nineteen Eighty-Four*'s Ministry of Truth to pull off such an operation. He got newspapers and publishing houses to do it themselves. As the late historian and archivist David King wrote in his excellent book, *The Commissar Vanishes, The Falsification of Photographs and Art in Stalin's Russia*:[105] 'photographic manipulation worked very much on an ad hoc basis. Orders were followed, quietly. A word in an editor's ear or a discreet telephone conversation from a "higher authority" was sufficient to eliminate all further reference – visual or literal – to a victim, no matter how famous she or he had been.'

Yet, it wasn't just newspapers and publishing houses that altered photographs on Stalin's command. Ordinary Russians were required to as well. When one of their family members was 'disappeared', they were required to tear out or ink over that family member's face in personal photographs. The same process was required for any disappeared person featured in photographs in old magazines or newspapers a citizen still had a copy of.

Stalin didn't just opt to achieve historical negationism by simply removing traces of disappeared people from photographs, though. His photomanipulation operation also included having himself inserted into historical photographs at events he hadn't been present at. One such example is a 1922 photograph that ostensibly shows Soviet founder, hero and icon Vladimir Lenin and Stalin seated very close to each other, looking quite friendly, at a country estate near Moscow. In reality, Stalin was added to the photo later after Lenin's death, with the altered photo widely reproduced in the 1930s. When the original photo was taken, a progressively ill Lenin was largely concerned and wary of Stalin and his growing power.

If only Stalin's fellow party comrades had been so concerned. In another photo, originally taken in April 1925, Stalin stood in the centre of nine party comrades. When the photo was republished in an official Stalin biography in 1939, three-fifths of the people in the original photograph had been cropped or retouched out of it. This left Stalin and three other comrades, two of whom had been completely relocated from their position across the room in the original photograph to stand within arm's reach of their leader in the altered one.

Sometimes Stalin even had people removed from photos he most likely did not have 'disappeared' in real life. It's widely known that by the 1930s, Stalin had utter scorn for the ordinary worker, somewhat ironic considering the Soviet Union's version of communism alleged the ordinary worker was of the utmost importance, and Stalin paid constant lip service to the working class in his rise to power. This scorn is no more evident than in a manipulated photo originally taken at the Sixteenth Party Congress in 1930. The original photo showed Stalin standing on a curb, gazing out of frame. To his right was a common worker pointing in the direction of Stalin's gaze. In other words, the worker was helpfully giving the great Stalin directions to somewhere. When the photo was reprinted in a magazine shortly afterwards, as King points out, 'no humble worker was there to tell Stalin where to go'. This was not a simple crop job. The most advanced image manipulation techniques of the day were used to erase the worker as if he had never been there.

Stalin wasn't just fond of rewriting the past by inserting himself into photos or erasing people from photos he appeared in, he also had a penchant for forcing film directors to re-edit their films to remove references to historic figures. After Stalin saw an early screening of Sergei Eisenstein and Grigori Aleksandrov's *October: Ten Days That Shook the World*, he had the filmmakers re-edit the film to cut all references to his one-time comrade Leon Trotsky. Stalin also didn't bat an eye when photographic stills from the film came to pass as historical records of the actual October Revolution.

It's unknown how many photographs Stalin faked, re-contextualised or outright manipulated, but King alone uncovered thousands of examples. *Thousands*. And remember, this was the first half of the twentieth century and manual image manipulation was much more tedious than its digital counterpart today. Is it now any wonder Stalin's photographic falsification operation was a model for *Nineteen Eighty-Four*'s Ministry of Truth?

While Stalin couldn't have possibly conceived of technologies like digital video and artificial intelligence, imagine what he would have done with deepfake technology – with the ability to alter history at will and put any words you want in your self-declared public enemy's mouth? Given his proclivities for falsifying still photography, Stalin probably would have sacrificed all of East Germany, Belorussia and Ukraine for just one computer with DeepFaceLab on it. Deepfake technology would have enabled historical negationism on an undreamed-of scale in Stalin's Russia.

Panetta and Burgund created *In Event of Moon Disaster* to educate, inform and warn. Stalin, had he lived to the Space Race-era and had deepfake technology to hand, most probably would have created the exact same deepfake to lead his isolated citizens to believe that the Soviet Union's inferior enemy had botched the moon landing, killing two of its greatest heroes in the process. Now, what will North Korea's Kim Jong-un, Iran's Ali Khamenei or Russia's Vladimir Putin use deepfakes for?

In fact, we barely even have to imagine. Some leaders are already showing us.

In February 2020, Manoj Tiwari, President of the New Delhi state Bharatiya Janata Party (BJP), allegedly became the first politician to use deepfake technology in an election campaign that was not done for overtly satirical purposes, according to a report from *Vice*.[106] Tiwari's campaign allegedly released three different versions of the same video across social media and in WhatsApp groups. In the original video, Tiwari spoke in his native Hindi.[107] The other two versions of the video had Tiwari speaking in English[108] and Haryanvi,[109] a Hindi dialect spoken by voters the BJP had hoped to sway. The videos of Tiwari, speaking in two languages he is not known to speak, were allegedly created with the approval of the BJP in order to court urban voters and migrant worker populations in New Delhi, respectively. Voice actors spoke Tiwari's message and deepfake technology then synced his mouth's movements to the imitated speech. The social media manager for the BJP New Delhi party told *Vice* the deepfaked videos were shared in 5,800 WhatsApp groups, being viewed over 15 million times, boasting, 'Housewives in the group said it was heartening to watch our leader speak my language'.

While the ability to share information in multiple languages is useful and, in many cases, a noble, admirable endeavour, it's not hard to conclude that many viewers were probably unaware that Tiwari wasn't really speaking their language. And who can blame them? The deepfakes do not appear to feature any disclaimers that Tiwari's voice, language and facial features such as lip movements had been synthetically manipulated. After uproar spread about the deepfakes, a BJP spokesperson disavowed the videos, saying the Haryanvi deepfake was created without their consent and then sent to the party.[110] Wherever the truth lies, it's charitable to say those who cast their vote on the belief that Tiwari really could 'speak their language' were misled.

This is not the only example. There's the military coup that almost happened in Central Africa simply due to the awareness of deepfake technology.[111]

In October 2018, local media in Gabon reported that Ali Bongo, the country's president, had been rushed to the hospital. In the weeks that followed, little news came out about his status. For a nation accustomed to regular appearances from their leader, the stark absence was abnormal to say the least and soon rumours flourished online that he had died, and the government was hiding his death from the Gabonese people because Gabonese law states that when a serving president becomes unfit, infirm, or dies, a special election must be held within sixty days – putting the ruling party's power at risk. With no sign of Bongo and rumours of his death swirling, in December 2018, the country's vice president revealed that Bongo had had a stroke but was recovering at a private residence. Still, that did not quell the rumours. Why else had he not been seen since his supposed move to recovery?

Then on 1 January 2019, Ali Bongo made his first appearance in almost ten weeks.[112] A video of the president was posted to his social media channels,[113] but, the three-minute video in which Bongo admitted he had been through a 'difficult period' only raised more suspicions. The video seemed *off* to many. Bongo's eyes and face appeared to be motionless at times. The voice and speech patterns didn't seem right. His facial expressions didn't seem cohesive, as if his jaw and his eyes had been Frankensteined together. All this may have been somewhat explained by the natural consequences of having suffered a stroke, but there was also the issue of blinking. The man in the video blinks only thirteen times in the two minutes and nine seconds he is featured. Most people blink close to forty times in two minutes. Back in late 2018, one of the big tells a video was deepfaked was that a synthetic face didn't blink, for the simple reason that the AI was trained on images and videos of people who had their eyes open (as most of us do in photographs), and so had no knowledge of the fact that humans blink regularly. (Deepfake algorithms have since learned about blinking and corrected for this.)

All this seemed to add up to the fact that the video was a deepfake, with many positing that the government was merely using it as a pretext to avoid holding a special election, and on 7 January, the Gabonese military attempted a coup against Bongo's government. It was the first attempted coup in Gabon since 1964. It ultimately failed and, in August 2019, Bongo made his first public appearance. He remains the president of Gabon, at the time of this writing.

The truth of the video is unclear to this day. Some claim that the odd speech and expressions in the video back up the claims that Bongo did, in fact, have a stroke. Others maintain that while Bongo did have a stroke, he was not the man who appeared in the video, that the government was merely stalling for time. Whatever is true, the fact that the public knew deepfake technology existed, and those who wanted to believe the video was false could point to deepfakes to explain how the video was created, is partly what enabled and emboldened the Gabonese military to attempt their coup. Their pretext was the video was not authentic. The challenge, how could someone fake a video like that? The answer, deepfakes – *duh*.

The Bongo 'deepfake' incident shows the power deepfakes have over societies just as a mere concept. Now that deepfakes exist, those who see something that does not agree with their agenda can allege the video is a deepfake even if it's not. We're all too familiar now with claims of 'fake news', yet how chilling is it that true fake news could be indistinguishable from real news? That the mere suggestion of falsehood may be enough to undermine the truth? Such denials could radically destabilise the trust in democracy – and yes, even be the pretext for coups.

And militaries aren't the only branches of state-authorised enforcement using deepfakes (or just the concept of them) to their advantage. As we've seen already, the use of deepfakes is not always bad news – though this next instance is sobering indeed. In early 2021, German lawmakers gave law enforcement sweeping new powers allowing undercover police to use deepfakes to infiltrate child sexual abuse gangs.[114]

In recent years, much organised criminal activity on the internet has migrated to what is known as the 'dark web', where you can find forums that allow you to buy or trade drugs, malware, illegal pornography and even find assassins for hire. As far as illegal pornography goes, on the dark web, this can come in the form of dedicated forums that trade in sexual abuse images of children. To access these forums, users must first 'prove themselves', for lack of a better phrase, by providing images of child sexual abuse. In Germany, it is usually illegal for law enforcement to use actual child sexual abuse imagery to catch criminals, thus, dark web forum owners know if people can supply such images, they are likely not law enforcement.

Germany's new law enables law enforcement there to use deepfake technology to generate images featuring children that never existed in real life. These images look 100% real because the deepfake software used to create them has been trained on a database of actual child sexual abuse images. That doesn't break existing German law as the deepfaked images are 100% synthesised. In this way, the deepfaked child sexual abuse images grant law enforcement access to criminal underworlds they would have no other way into, meaning they can monitor, track and catch perpetrators more easily.

Though this is a major coup in terms of end result, the fact that deepfakes are being used to generate photo-realistic sexual images of children has understandably raised concerns. Specifically, it's that law enforcement agencies are now actively contributing to the creation and spread of new pornographic images of children on the dark web, images that would not have existed otherwise had they not deepfaked them. Thankfully, there are restrictions on the use of deepfakes in this manner. Only synthetic people are allowed to appear in the deepfakes and authorities using this tactic to gain access to dark web forums first need a judge to grant permission. They also need to participate in special training beforehand. However, despite these checks and balances, I don't think one would be wrong in saying this use of deepfakes sits on the very edges of what most of us might deem acceptable methods. The ends well may justify the means but that doesn't mean it's a particularly palatable option.

This also opens up the question if German law enforcement is using deepfakes to target and catch one group of people (criminals), what other groups might law enforcement agencies in other countries with less liberal democracies and fewer checks and balances use deepfake technology to ensnare?

In China, where media is censored, could a deepfake of a party member asserting democratic values and encouraging others to do the same be used to weed out political dissidents?

In Chechnya, where homosexuality is openly persecuted, could a call to a human rights meeting for like-minded individuals from, ostensibly, a known gay-rights activist be used to entrap and round up members of the LGBTQ+ community?

In Myanmar, which was taken over by coup in February 2021, could the ruling military create a deepfake of a beloved pro-democracy protester urging others to join her on the streets at a certain time and place – only for the military to be waiting with their guns ready to fire?

The answer to all of these questions is 'yes'.

And if some political leaders and governments are already using deepfake technology to their advantage, is it any wonder that other governments are preparing for a day when deepfakes may become the preferred tool of geopolitical adversaries? It's not hard to imagine a day where nation-states don't just use deepfakes against their own citizens but against the citizens of their enemies via massive and sustained disinformation campaigns designed to sow confusion or anger among a country's population. Propaganda has potentially never been so dangerous.

In late 2015, a DARPA program launched called MediFor.[115] Short for 'media forensics', it was created with the aim of inventing automated tools that can assess the integrity of images or videos. But as we've already seen, even if you can prove a video is a deepfake after the fact, sometimes the damage has already been done.

To help mitigate this, in December 2019, the US Congress passed the first federal deepfake legislation.[116] Unlike previous state legislations we've already discussed, the federal legislation was solely concerned with the weaponisation of deepfakes by foreign adversaries. The legislation is part of a $738 billion defence policy bill and mandates that the Director of National

Intelligence must make Congress aware on an annual basis of any potential national security threats from deepfakes. Specifically, the bill requires appraisals of how China and Russia already are or may use deepfake technology in the future to disseminate misinformation, discredit disfavoured populations or political opponents or otherwise harm the United States or its allies.

Furthermore, the bill authorises the establishment of a 'Deepfakes Prize Competition' to award cash prizes to researchers who build technologies that can automatically and accurately detect deepfakes. Up to $5 million can be awarded to one or more winners. The bill was passed with overwhelming cross-party support in the House and Senate and then signed into law on 20 December 2019 by then-president Donald Trump.

If you needed any more persuasion that deepfakes are a threat to be taken seriously, this is it.

Yes, it's true, as we've seen, deepfakes have several positive use-cases – from entertainment to art to healthcare to fashion to education. Yet deepfakes can also be used to bully, abuse, harass, humiliate, disinform, scam, extort and now convincingly alter history. It can be used to show any one of us – celebrities, politicians, the girl next door, me, *you* – any one of us saying and doing things we never did.

So what happens in a world where we can no longer be sure if what we see on our screens is true or false? If we can no longer believe what we see with our own eyes and hear with our own ears? Deepfakes could create a zero-trust society within the next decade, and that's something we all need to be prepared for – and understand what we can do to fight back.

Chapter Seven
The End of Trust

In January 2019, a new user joined Twitter under the handle @Azusagakuyuki. As with most new Twitter accounts, @Azusagakuyuki did not post much in the beginning. Slowly, however, the user began to tweet more and more about one subject: motorcycles. Turns out @Azusagakuyuki loved riding motorcycles across Japan and would frequently tweet and reply to a small group of followers and fellow enthusiasts, trading tips and talk about the pastime. Yet most of @Azusagakuyuki's tweets never got much engagement. A few replies here and there, a few likes – maybe a dozen on the most popular posts.

But by mid-2020, something changed and @Azusagakuyuki began not just tweeting thoughts about motorcycling but tweeting pictures as well. Most of these pictures were of the motorcycle @Azusagakuyuki rode – a red and white Yamaha sportbike. In some of these posts, though, @Azusagakuyuki featured in the picture, too.[117] One of the early photos featuring @Azusagakuyuki shows the motorcycle enthusiast in front of the Yamaha being worked on. The motorcyclist gives the peace sign and states that the sportbike's cylinder is in mid-repair due to a leak from a stud bolt. The picture reveals the red and white bike partially disassembled, its fuel tank covering removed. It also reveals another thing: @Azusagakuyuki is a woman, and a beautiful one at that. She's Japanese, which you could always tell from the language she tweeted in, and has gorgeous, long caramel-brunette hair and a friendly smile. She looks maybe twenty-five at best.

The photo tweet in question garnered over 120 'likes'[118] – that's way more engagement than @Azusagakuyuki ever received before when simply tweeting about motorcycling without any pictures of herself attached. So, as time went on, @Azusagakuyuki did what any other person on Twitter would do when they discovered the types of posts that drive more engagement. She still tweeted about her favourite pastime, but now she frequently included pictures of herself along with her thoughts on the sport. Sometimes the entire tweet would just be a close-up selfie of @Azusagakuyuki, but many times the photos would take the form of a wide shot of @Azusagakuyuki with her red and white sportbike, perhaps on the road or working on it in her apartment.

As the frequency of the tweets featuring herself in the photos continued to increase, so too did the likes @Azusagakuyuki's posts received, with her follower count growing at a commensurate rate. And it's obvious why: here was a woman who was young, pretty, knew her way around a toolbox and could ride and talk bikes with the best of them. In other words, @Azusagakuyuki was the 'hot biker chick' most male motorcycling enthusiasts dream about. Unsurprisingly, throughout 2020 and into 2021, @Azusagakuyuki's follower account continued to grow, as did the rate she tweeted photos of herself, and by February 2021, @Azusagakuyuki had over 20,000 followers.

Then on 11 February 2021, @Azusagakuyuki tweeted another set of photos.[119] This one included a profile of her Yamaha motorcycle inside her apartment; another snapshot of the

Yamaha shot from the rear, so you see its rear wheel, then its seat, and finally handlebars and side mirrors; and the final photo of the set was a snap of @Azusagakuyuki herself, looking quite adorable, seated at a small workbench, smiling, and surrounded by tools like pliers and wire strippers.

Along with the photos was the message that she was working on her bike all day, making adjustments to the amplifier and running wiring. The crying emoji she uses after this suggests it was tough work. But she shrugs it off, ending the message with another cute kawaii text-based emoji.

Just another ordinary tweet from the hot biker chick, right?

From first appearances, yes. But then one of her followers noticed something. In the second photo in the set – the one of the Yamaha shot from the rear – there was the glimpse of a figure in the motorcycle's left and right rearview mirrors. The left mirror showed the body of the figure – it was the person taking the photo. You could only see them from the chest down, though. They wore a blue-grey shirt and held their phone's camera pointed towards the sportbike. The right mirror, however, revealed more due to its vantage point. This mirror showed the face of the figure taking the photo. It was a bit fuzzy, but if you zoomed in enough, something caught your eye. The figure reflected in the mirror didn't look like @Azusagakuyuki at all. The person was wearing eyeglasses and kind of looked like, well, a man.

Whoever took this photo was not @Azusagakuyuki.

Was it a boyfriend – someone she'd never mentioned or tweeted about? After all, it's not uncommon for some influencers to refrain from mentioning their partners or romantic lives. In some instances, this is to maintain a clear line between their public persona and their right to personal privacy. In other instances, this is just a shrewd branding move. Some female influencers, particularly, have been reported to get less engagement from male followers if their relationship status is already known to be 'taken'.[120] It's why some influencers remove their wedding ring before taking photos to be posted to social media. Could one of these situations explain @Azusagakuyuki's unidentified man?

Perhaps . . .

'Perhaps' wasn't good enough for the producers of a popular Japanese variety show called *Getsuyou Kara Yofukashi*, however (the name roughly translates to 'Monday Late Show'). The producers sent a team out to Ibaraki, a prefecture north-east of Tokyo that borders the Pacific Ocean, to track down the increasingly popular and mysterious @Azusagakuyuki.[121] But when they found her, they were not confronted with a young woman with that beautiful face and gorgeous, long caramel-brunette hair. Instead, the @Azusagakuyuki they found took the form of a fifty-year-old man. His name was Soya and he confessed to the team from *Getsuyou Kara Yofukashi* that he was the hot biker chick everyone loved. For nearly a year, he had been using the popular smartphone deepfaking app FaceApp to change his age and gender in the selfies he took to post to Twitter.

The one saving grace for @Azusagakuyuki's admirers, those who had fallen in love with @Azusagakuyuki for 'her' gorgeous, long caramel-brunette hair anyway, is that it was at least real. Those long locks are genuine, still gorgeous and actually attached to Soya's head. Only his face's age and gender were deepfaked using FaceApp. After the team from *Getsuyou Kara Yofukashi* tracked Soya down, he agreed to 'come out' on the popular show.

'No one will read what a normal middle-aged man, taking care of his motorcycle and taking pictures outside, posts on his account,' Soya said on the show, noting that he believed he could get more social media engagement as a 'younger beautiful woman' rather than an old 'uncle'.[122] That's when he thought of FaceApp.

'First, I just tried [deepfaking my face to become a woman], then it happened to turn out to be fairly pretty. I get as many as a thousand "likes" now, though it was usually below ten before.' He added, 'If a girl like her existed, anybody would go crazy for her, right?'[123]

Based on @Azusagakuyuki's skyrocketing follower count once 'she' started posting selfies, yeah, Soya was spot on.

Many of @Azusagakuyuki's followers took the reveal in their stride, some commenting that they outright loved the unanticipated turn of events. Some told Soya his real face and his deepfaked face were both beautiful. Others said no matter if he were a man or woman, they were still jealous of Soya's gorgeous head of hair. However, some followers admitted that they didn't think they would ever fully trust someone's identity on social media ever again. Soya, they said, had catfished them.

To these followers, others countered that it wasn't really cat-fishing at all. Soya's defenders pointed out that Soya didn't fake his love of motorcycling, his advice to others or his skills on the road – which is what @Azusagakuyuki's fans claimed they were following 'her' for to begin with. And Soya didn't have any nefarious intent. You could argue that Soya's gender is – or should have been – immaterial to the content he was posting. All he changed was his face; the overall context of the photos remained the same.

So, was what Soya did wrong? I think it's hard to argue that it was – at least in any significant way. After all, he wasn't trying to meet up with other men or women in real life by pretending to be a beautiful young woman, nor was he selling products based on his fabricated facade. On the contrary, his actions were relatively innocuous. As some of his fans pointed out, besides his deepfaked face, everything else in the pictures he tweeted was real. It was really him riding the Yamaha down a coastal road or working on it in his apartment. It was his words and thoughts that he tweeted when talking shop with other motorcycling fans, many times giving valuable advice to newbies who recently picked up the pastime. At the end of the day, Soya wasn't some kind of malicious actor. He was just a crafty middle-aged man who seems to genuinely love discussing all things motorcycles and who just wanted more followers and online engagement for his thoughts.

However, this viewpoint only holds up because we are examining Soya's actions as an isolated incident. He deepfaked himself, no one suffered demonstrable harm from it and he admitted it when confronted. Not so bad, right?

Yet what happens if we zoom our perspective out from just Ibaraki Prefecture to all of Honshu? Then from there, let's zoom out to all of Japan, then all of Southeast Asia. Now let's zoom out further to all of the Eastern hemisphere, and then zoom out more to include all of the Western hemisphere, too, and keep going until the whole planet is within our perspective.

From this global perspective, what happens when we see that there are ten thousand other Soyas doing something similar? Or a million? Or ten million? And what happens when you add those incidents to all the other deepfaked incidents across the globe? When you add the other deepfaked social media profiles together with the fun face swap selfies posted everywhere and the entertaining deepfaked YouTube fancasting videos? Then add those in with the deepfaked celebrity porn and the deepfaked revenge porn videos. Now sweep all that up with the deepfakes created to extort – the videos making it look like someone committed a crime they didn't do – or the deepfaked audio that makes you think it's your boss who's on the phone and needs the money wired. Keep going. Let's add in the deepfaked virtual influencers designed to engage you with a brand's products into the mix, and the deepfaked actors designed to allow the billion-dollar cinematic universe to continue long after their deaths. And let's not forget the deepfaked political ads designed to influence your vote, and the educational deepfakes of historical events like the moon landing designed to bring history alive, and the law enforce-ment-approved deepfake pornography designed to catch child predators and, of course, the deepfakes designed to destroy not only political campaigns, but entire nations.

Examining Soya's actions as just one among a multitude of other similar incidents, we start to realise the ultimate impact deepfakes will have. It doesn't matter whether the individual deepfake aims to make you laugh, get you off, humiliate an ex, sell you something, educate you, entertain you, entrap you or bring down a politician, party or government. It's an impact that is only realised when the effects of individual deepfakes are looked at cumulatively. And that cumulative impact is orders of magnitude greater than the sum of its parts. It's an impact some of @Azusagakuyuki's dejected followers alluded to – yet it's enlarged from the individual to the societal level. That impact is one of the erosion of trust.

We will soon live in a world where we will need to ask of everything, 'is that real?' because we will no longer be able to trust that the photos we see, the videos we watch and the audio we hear are authentic representations of fact. Even now, with every few months that pass – just *months* – deepfake algorithms only become more polished and powerful. With every year, the tools used to create photorealistic deepfake videos will continue to be refined and simplified until anyone with zero knowledge about GANs or audiovisual manipulation will be able to make a deepfake of anyone else with a few clicks of a button.

As hard as it is to believe, everything we've already explored, from DeepFaceLab-created fake celebrity porn videos to Soya's @Azusagakuyuki deepfaked Twitter persona, are just the first drops in the greater tsunami of synthetic change now barrelling towards our media shores. At some point in the near future, the majority of audiovisual content that we find online – and even on television and cinema screens, if those mediums last that long – will have been synthetically altered in whole or in part by artificial intelligence. Yes, that synthetic alteration – that deepfaking – may bring joy and delight. It may allow new Beatles songs to be created as if John and George never left us. It may enable us to see Carrie Fisher playing Princess Leia once again – forever. Yet, via our social media channels and our online bubbles, especially, it may also bring outrage and indignation. Fear and loathing. It will allow anyone to craft any narrative they want – and back it up with audiovisual 'evidence'.

No, Donald Trump was the greatest orator in the history of America. Here's the video that proves it.

Barack Obama was nothing more than a mumbling, bigoted fool who couldn't string enough words together to form a coherent sentence. See here.

Check out my kid scoring the winning home run. No, no. It's not a deepfake. It's real. My kid is obviously better at baseball than my brother's.

She said I called her a what? Listen to this audio I recorded of her at my door threatening to kill me.

I liked Becky as a candidate, too, until I saw this old Facebook video of her using racist language back in her college days.

No, that's the real speech Nixon gave when the astronauts died on the moon. The clips of them bouncing around in low gravity are the deepfakes.

I really am a hot biker chick. Look at all the videos I've posted of myself blasting the pavement in Hokota on my crotch rocket. You believe me, don't you?

Don't you?

Harmless or malicious, deepfakes will never again allow us the certainty that we once took for granted – the foregone conclusion that we can trust what we see with our own eyes and hear with our own ears.

We will soon be living in a world of unreality.

The resulting erosion of trust that comes with this, which is so fundamental to social cohesion, will almost certainly have grave consequences for our wider world. For what is trust if not the binding material of a society? Without trust, there is no cooperation. Without trust, the institutions society depends on to operate risk collapse – governance, the justice system, scientific bodies and educational institutions, the press and news media and even the military.

At best, this coming synthetic tsunami, this erosion of trust, could result in what scholar and non-resident fellow at the

Alliance for Securing Democracy, Aviv Ovadya, calls 'reality apathy'.[124] Simply put, people may give up trying to *determine* what is real and what is fake. There's just too much manufactured interference to separate the signal from the noise. It's too exhausting to try, and you'll never know if the conclusion you come to is, in fact, the right one – so why even make an effort? But we also run the risk of an epidemic of cynicism as we come to assume – out of necessity – that everything we see or hear is fake. If nothing is real, nothing matters.

The thing is – this widespread reality apathy and cynicism, while it's clear to see how detrimental it will be to society, may allow some individuals to thrive. That's because of a paradox inherent in the existence of deepfake technology: though you may be able to prove a video *is* a deepfake, you can never prove with absolute certainty that a video *is not* a deepfake. This paradox allows anyone to dismiss authentic recorded events as being falsified. And if enough of society is so cynical or reality apathetic, they may let those claims of falsification slide, which is why this concept is known as 'the liar's dividend', a term coined by law professors Danielle Citron and Robert Chesney.[125] The dividend here is that liars who actually get caught on video or audio doing something wrong can now plausibly dismiss the underlying audiovisual evidence. Or taken a step further: anyone can now claim that something is faked – to the benefit of their narrative. This isn't even too far from the reality that we've just experienced, and we all know when the lies start to pile up we become more apathetic as a society. We let more things go. Imagine that, all the time, with everything.

Imagine a politician is caught on video saying something socially offensive, she simply needs to reply: '*It was a deepfake.*'

When a spouse goes to the police with a voicemail from her partner, threatening to harm her, now her partner can claim: '*That's a deepfake.*'

When confronted with iPhone footage showing a drunken man smashing car windows after leaving the stadium where his favourite team just lost, the man says: '*Duh. Deepfake.*'

These are just hypothetical examples of the liar's dividend. Want real-world ones?

A month after the video of the murder of George Floyd at the hands of police in Minneapolis went viral in May 2020, spurring the Black Lives Matter movement to new heights, a right-wing candidate running for a Missouri Congressional seat published a 23-page report on Floyd's death. That report alleged that the 'real' Floyd died in 2016, and thus the Black man in the May 2020 video with the police officer on his neck was, well, fake.[126] A deepfaked-generated man, to be precise. Oh, and so were the police officers in the video.

'We conclude that no one in the video is really one person but rather they are all digital composites of two or more real people to form completely new digital persons using deepfake technology,' the report read, adding, 'We urge President Trump to open an investigation into these claims to help resolve the issue of deepfake technology once and for all.'

Far from sounding like a crackpot's ramblings, the existence of deepfake technology means that these claims could, in the right (or wrong) circumstances, sound like the voice of reason.

Thankfully, the candidate arguing for the Floyd deepfake theory lost the Republican primary several months later.

We need to only wait a few more months for another example of the liar's dividend in action. In October 2020, then-President Trump was rushed to the Walter Reed Military Hospital after being diagnosed with Covid-19. The diagnosis was not only a hard thing for Trump to deal with (personally and politically), it was hard for his followers as well – many of whom believed in conspiracy theories that the global pandemic was a made-up hoax all along. These were theories that Trump often failed to repudiate, knowing how popular they were among his base. So how did these Trump followers reconcile the fact that their strongman had caught the 'fake' Covid-19 disease – especially after footage of Trump going to, staying at and ultimately being discharged from the hospital was broadcast around the world over several tense days?

Some simply didn't. Instead, they claimed that video and audio clips of Trump announcing his Covid-19 diagnosis were obviously deepfaked. Though their claims probably convinced very few people outside of their Facebook groups and Parler threads, it did allow them to continue to ignore reality, to lie to themselves and dismiss the unthinkable away – and respond to non-believers that they couldn't definitively prove otherwise. This same argument was used again on 7 January 2021, a day after insurrectionists stormed the US Capitol. The White House posted a pre-recorded video of then-President Trump finally condemning the previous day's violence. Many on Qanon forums alleged the video was a deepfake created by Trump's enemies, and thus the president was still on the insurrectionists' side.

For another example of the benefits of the liar's dividend in action, we simply need to return to something we already discussed. Remember the attempted coup in Gabon? The liar's dividend is what allowed some in the Gabonese military to assert that the 1 January 2019 video of Ali Bongo was a fake, that the president was really incapacitated or dead and that the government was trying to fool the people otherwise. In my opinion it's unlikely the Gabonese military actually believed this, but the liar's dividend gave them a needed pretext that would not have been feasible just years earlier. The existence of deepfake technology made their claim possible, if not too plausible, given the development of the technology at the time. But then again, the liar's dividend works precisely because it relies solely on possibilities – the possibility that anything can be a deepfake, and you can't categorically prove it's not.

One final thing to mention about the liar's dividend is that some experts believe its potential to dismiss the obvious truth with a straight face could be a boon to China,[127] which passed the first national anti-deepfake law in the world. On 1 January 2020, new rules came into force in the People's Republic of China that banned the publishing and distribution of 'fake news' created with artificial intelligence. At a press briefing announcing the new rules, officials from the Cyberspace Administration of China stated that the law was needed because deepfake technology could 'endanger national security, disrupt social stability, disrupt social order and infringe upon the legitimate rights and interests of others'.[128]

All that is true. Malicious use of deepfake technology has the potential to cause such mayhem. However, it's not hard to theorise that China's anti-deepfake law was more than just about stopping the spread of online disinformation and social disorder in the country. Since 'fake news' deepfakes are now illegal in China and cannot be posted or shared, the Chinese Communist Party has a powerful new censorship tool in their arsenal in the form of the liar's dividend. The CCP can now label any audio-visual content detrimental to the party or social order – even if it's authentic – as a deepfake without cracking a smile.

Has footage emerged of the country engaging in environmental, war, human or civil rights abuses? 'It's a deepfake,' say straight-faced CCP officials. 'And if you post or share it, you're a criminal.' And if the CCP says it's a deepfake, who is a Chinese citizen to argue – especially if they have any sense of self-preservation? The liar's dividend is the perfect censorship pretext for dictators and authoritarians. Stalin could only wish he'd had it so good.

Over the next decade, we'll see the maturity of artificial intelligence – and the best computers will move from the silicon-based world to the quantum world, allowing for computational processes to be carried out at undreamed-of speed and power thanks to quantum computing. These quantum computers will only be in the hands of the biggest tech giants and most advanced nation-states. And they will enable realistic deepfakes to be created in – not days – but hours, maybe minutes.

Right now, even the best deepfake AI can only manipulate already existing video. By the end of this decade I'd need only instruct a quantum deepfake AI to create a video of George Clooney sitting in a Starbucks drinking a latte with Barack Obama discussing if Michael Grothaus or Leo Tolstoy is the better writer, and it will fabricate the photorealistic synthetic scene out of thin air, including the people, tables, coffee mugs, lighting, conversation and background sounds.

It sounds like magic, I know. But in 2000, computers in our pockets with multitouch screens were equally fantastical. Just eight years later, they were commonplace. In 2010, AI-manipulated video seemed like science fiction. Four years later, GANs made it science fact. In 2021, already several startups are developing camera-free synthetic video deep learning systems that will allow for the spontaneous scene generation, like the kind needed to create a record of Clooney and Obama debating my writing abilities in that Starbucks – and these deep learning systems can run on the computers consumers like you and me have access to. Once such camera-free synthetic video creation is sufficiently advanced, you have to wonder if we will even have an ounce of trust left in any media we see on our screens?

If you're feeling unnerved at this point, I'm sorry. I am too – we're in this together. I'd love to be able to say at this point that there's a way out, but the truth is, we can't turn back the tide. But we can prepare. We can mitigate.

First, we shouldn't only rely on legislation. Yes, some legislation may be appropriate, like amending revenge porn laws to restrict the creation and sharing of non-consensual pornographic deepfakes of others. This would help victims who would have trouble bringing civil or criminal charges against their pornographic deepfake creators on existing harassment or defamation grounds. However, countries – democracies anyway – must be careful not to tread on freedom of speech guarantees. As a tool for free and critical expression, deepfakes have profound value. The technology can be applied to create powerful social commentaries and political satire of public figures and events, just as political cartoons have done for centuries and comedy sketch shows have done for decades. We must ensure that kind of speech is always allowed in a free society, no matter the tools used to generate it.

As for legislation criminalising deepfake technology itself? That would be a dangerous path to go down (what technology do you outlaw next?). It would also be pointless. After all, most drugs are illegal, yet the drug dealers just don't seem to care. Deepfake technology is already out there – it's just digital code that can be transferred around the world to any computer in a matter of seconds. Besides, you're not going to stop the people who want to use it for nefarious purposes from doing so just because a law says they can't.

Instead, mitigation efforts should be focused on two fronts: detection and education.

Detection is the most obvious – it's also the most problematic. Almost since the GAN technology used to create deepfakes came into being in 2014, there have been attempts to create tools to identify if a video has been deepfaked. These detection tools, of course, are powered by artificial intelligence, just like the tools used to create deepfakes. Many of these deepfake detection tools learn in the same way as deepfake AI does. The detection tools view datasets of real videos and deepfake videos and then use that data to label a video as legitimate or a deepfake.

The problem with this approach is that you are essentially pitting two groups of AIs against one another – deepfake AIs and deepfake detection AIs. When the deepfake detection AIs find a new 'tell' – a new giveaway that reveals a deepfake is a deepfake, the deepfake AIs soon learn to correct for this so they can make improved deepfakes that can fool the deepfake detection software. We've seen it before when AI learned to correct for blinking. From here, it's only a matter of time before deepfake AI learns to correct for the next thing that could identify a deepfake as a deepfake – like learning to generate reflections from the surrounding environment in the subject's eyes.[129]

Ultimately, this kind of mitigation is the AI equivalent of an eternal cat and mouse game. Sometimes the deepfake detection tools will be ahead; other times, the deepfake creation tools will be ahead without the deepfake detection tools even realising it. It's a game that will never end. Of course, that's not to say that deepfake detection tools are pointless. Even though they won't detect every deepfake, the fact that their AI continually gets better at detecting deepfakes means the bar is raised for deepfake creators everywhere. Because of deepfake detection tools, a deepfake has to be of at least a certain quality to even have a hope of fooling detection software. This ever-rising bar means it will take longer to create a high-quality deepfake, and that the computers used to create deepfakes that can pass that bar may not be accessible to today's average deepfake creator who relies on a PC with off-the-shelf parts. This constant raising of the bar is why government agencies like DARPA and tech giants like Microsoft, Google and Facebook – and thus society – will see dividends from such detection tools they are developing even if the tools don't have a perfect success rate.

Besides AI detection tools, there are other technology-based detection methods that we can use to identify deepfakes. Some possible methods could include blockchain timestamping.[130] This would assign a cryptographic hash – essentially a digital fingerprint – to every piece of digital content created and store this fingerprint and its timestamp on the blockchain. The timestamp would show when the digital content – a video of a politician, for example – was first created. If that video is then altered in any way in the future, that altered copy would automatically receive a new fingerprint, uploaded to the blockchain, which would identify the fact that the original source content has been manipulated and when. While any deepfakes would get their own fingerprints too, what blockchain time-stamping enables is a way for researchers to track a video back throughout time, identifying earlier versions and allowing any changes to be compared. The earliest version of the video is more likely to be the authentic one.

A method in a similar vein as blockchain timestamping is known as provenance-based capture.[131] Under this method, the device used to capture the original photo or video, such as a smartphone, could embed a digital signature into the media itself. That signature would contain information that is mathematically impossible to alter. Such information in the signature would include data about the device the photo or video was recorded on; the date, time and location of the recording; the pixel density; perhaps even data like the weather at the time the media was captured or facial recognition identification – so you know who originally appeared in it. If the content were then used in a deepfake later, the original data embedded in its signature may no longer match the synthetically altered copy, tipping the viewer off (well, the AI detector, let's be honest) that the content has been tampered with.

Blockchain timestamping and provenance-based capture aren't perfect solutions because they don't stop the deepfaking of the content to begin with, but they do offer a way for viewers, journalists, and government agencies to verify if the content they are seeing has been altered in some way. Platforms like Twitter, Facebook and the social media channels of the future could then automatically read the blockchain timestamps or digital signatures of content uploaded to its site and automatically flag or reject uploads whose timestamps or signatures show the media has been altered.

However, both of these methods will only be truly useful if adopted at scale by device manufacturers, social media platforms, messaging and other apps, websites, industry bodies and national governments. For example, suppose only ten per cent of smartphones support provenance-based capture, or only twenty per cent of websites and apps allow for users to quickly verify blockchain timestamps or digital signatures of the content displayed. In that case, you'll still have a wealth of places on the web where synthetically altered footage can be shared and posted without signs that something in the media has been manipulated.

It's these drawbacks to blockchain timestamping and provenance-based capture – and the cat and mouse nature of deepfake detection tools – that are precisely why we can't solely rely on automated deepfake detection tools to save us. That's why the most impactful mitigation techniques will not rely on detection at all, but education.

Deepfakes are not going away. And even though we may see a deepfake and viscerally feel that it's somehow *unnatural* for a video to be able to be manipulated to show something that, in fact, never occurred, it's actually nothing of the sort.

For centuries, if something was written in ink, it was fixed forever. It couldn't be changed once pen had been laid down to paper. We knew this. *That's how ink works*. But then, in 1979, Papermate invented the Erasemate erasable ballpoint pen and, with it, a medium that had always been fixed became flexible overnight. We just happened to grow up in an era where, until now, video and audio weren't malleable like other mediums. That is, the images and sounds that they recorded could not be changed at will later on – at least not to a realistic degree – nor could events that had not happened easily exist in the audio, photographic, or video record. That era – that aberration – has now passed. Mediums change, their capabilities expand, and we need to learn to adapt. Going forward, people will need to be taught that what they witness in audiovisual form isn't etched in stone. It could be easily fabricated, just like a false email, a fake text message or a Photoshopped picture.

This type of digital literacy, of digital awareness, will be crucial in fighting the psychological and social impacts the flood of synthetic media coming to our shores over the next decade will have. Such literacy should be started at an early age, with children. If you start early, you have a better chance at training individuals not to automatically believe everything they see on a screen.

This training should be complemented with a mandate that media containing synthetic manipulation – be it a film, a commercial, a podcast or, crucially, a political ad – must be transparent about the synthetic alteration. Users must be made aware the content they are watching has been synthetically manipulated. Political ads should be required to be even more forthright, for example, needing to explicitly state precisely what in the video has been synthetically manipulated (such as the politician speaking a language she doesn't actually speak).

Of course, this kind of synthetic media manipulation transparency will only partially help. Commercials and Hollywood films aren't likely to contain the malicious kind of synthetic manipulation that aims to deceive populations in nefarious ways. It's the deepfake videos that propagate across social media and in messaging apps that proclaim to be authentically-recorded events that are likely the ones to cause the most harm – the deepfakes made by bad actors with an intent to sow distrust and disinformation. If harm is the aim for the deepfake you create, you're not going to bother throwing a synthetic media manipulation transparency warning on the video no matter what regulations say.

This is precisely why education matters so much – to condition us in such a way that we almost automatically manifest a critical thinking barrier aimed at blunting the impact and ability to fool that nefarious deepfakes possess. That education should adopt teaching people to use media literary frameworks like SIFT, a methodology created by Mike Caulfield, director of blended and networked learning at Washington State University Vancouver.[132] Its aim is to prime people to pause and think critically about media before sharing or acting upon it. The FBI itself recommends using the SIFT methodology to help mitigate the impact of nefarious synthetic media campaigns[133] – in other words, deepfakes. Though SIFT's methodology can be applied to any information you come across, the following shows how to use it when confronted with a possible deepfake:

- **Stop**: in other words – breathe. Don't impulsively share or act on what you are seeing. Nefarious deepfakes, like other types of disinformation, aim to create a strong emotional response in the viewer – shock, fear, anger, outrage. These types of emotions are what compel us to share the disinformation without thinking critically about it first.
- **Investigate**: check out the video's source. Is that source reliable? If it's on a website, posted in a Facebook group, or sent as a WhatsApp attachment from your uncle, ask if they are generally non-biased – or do they have an agenda or are easily duped? And remember, simply because a video shared with you has a CNN or NHK logo, it doesn't mean it's actually from those sources.
- **Find**: seek out other coverage that details the event in the video you have been sent or seen. Is there other coverage of it? If not, why? For example, if it's a video of President Biden declaring nuclear war on China, are other news organisations reporting this declaration?

- **Trace**: can you follow the video back to its origin? Not just who sent it to you or where you've seen it posted, but who originally recorded it. Where and when was it recorded? Do the details in the video (the time of day, the weather, the people present) match up with conditions as they would have been if the video is authentic? And if that origin is a person who claims to have been there live and recorded it themselves, ask yourself, is that plausible? Are they likely to have been in that location or had access to it? And if they say they personally captured the video of a drunk politician beating up someone's grandma, does it really make sense that the supposed source would sit there and record it instead of stepping in to help stop the attack?

Finally, we need to teach people to protect themselves. You are more likely to be deepfaked if there is an ample supply of audiovisual media featuring yourself on the web. That's because deepfakes – for now at least – require large datasets on which to train if they want to be able to generate a lengthy, realistic synthetic copy of an individual. The necessity and availability of these large datasets are precisely why deepfake celebrity pornography and deepfaked YouTube fancasting videos have been among the most prevalent deepfake content created to date. The internet is filled with videos and photos of actresses and actors from which deepfake algorithms can train. Need to create a deepfake of a 22-year-old Anne Hathaway? Just download the 2006 film *The Devil Wears Prada* and then chop every second of the frames featuring Hathaway up into their thirty individual images and you'll have all the photos you need.

While most of us may not have a library of Hollywood films we've appeared in for deepfakers to draw from to train their models, most people who are now in their late teens and older have probably posted years', if not decades' worth of photos and videos of themselves to various social media platforms. This means there's more than enough video content online of the 'average' person today that a deepfaker intent on making a deepfake of them could get a hold of with little effort.

This even applies if you've only posted a couple of videos of yourself to platforms like Instagram or TikTok. When Brad originally asked me to record a video of myself so he could extract the individual images from the frames for my deepfake, he told me he only needed 60 to 90 seconds of footage of me – and he was playing it safe. Even one twenty-second selfie video would provide 600 photos – enough to train a model to make a realistic pornographic deepfake of yourself. And considering most people – especially those in their twenties and thirties – probably have combined *hours* of footage of themselves floating around their social media channels – well, a deepfaker has a pool of more than enough media to draw from in order to put you into their video doing anything they want you to.

That's why we're going to have to teach people who want to have the most protection to have better social media restraint. Every image or video you post can now just feed the deepfake AI tasked with creating fake videos of you. Some may want to consider restricting their social media posts to only family and very close friends (and always to only people you have actually met in real life). Ideally, people should refrain from posting any media of themselves if they wish to have the maximum protection against being deepfaked. If that seems extreme, unfair, even – admittedly, it is. But there's a reason we 'ordinary' people should be more concerned about deepfakes than the celebrities and politicians who already have their likeness spread wide across the web.

If someone makes a deepfake of Joe Biden or Julia Roberts shouting racist slurs or assaulting someone's grandmother, Biden and Roberts have the platforms, press officers and the financial resources to quickly issue statements far and wide that the video showing them engaged in unsavoury behaviour has been deepfaked. The rest of us don't have the luxury of having those resources at our disposal to spread the message rapidly and decisively that the video of 'us' is a deepfake.

Whether it's the armed robbery deepfake that was created of me or a pornographic deepfake of some non-famous victim, by the time we begin refuting the deepfake throughout our social circles to any significant degree, we could have already lost the trust of our friends, the standing in our community, our job. This is precisely why detection and education efforts are critical to mitigating deepfakes' harm to not just society but to individuals.

That being said, there is one truth that is important to acknowledge when it comes to deepfakes. And this is something no amount of detection tools or education can counteract. Even if you can prove to an individual, or even a large swathe of the population, that something they've seen is without a doubt a deepfake, that truth might not matter in the slightest. And it wouldn't be because those people have trust issues. It's because ultimately, we want to believe things that back up our existing beliefs. This 'confirmation bias' spurs us to seek out and believe information that confirms what we already think.

That very human tendency is why disinformation can be so effective. It's why obviously, ridiculously, fake news stories like Trump being endorsed for president by the Pope, which spread like wildfire on Facebook in 2016, and the almost just as obviously faked shallowfakes we've already discussed, like Skitz's 'The Hillary Song', work so well. It's not like these stories and shallowfakes are fabricated to such a high quality they are indistinguishable from fact. Rather, they work because they support what some people already want to believe.

The fact of the matter is, some people – a lot of us, it seems – don't appear to care if something might be fake if it backs up our existing worldview. Hell, even if people *know* something is a lie, sometimes they just want to be able to point to any evidence that supports their ignorance so they can shout, 'It's true! I'm right! *I've seen it with my own eyes and you can't tell me I didn't!*' to shut others on the opposite side of the argument up.

And even when not ideologically driven, our acceptance of the unreal abounds. Remember the virtual influencers? Many of Imma Gram's followers know she's a digital creation, but that doesn't stop them from following her, giving her compliments – or their sympathies, on occasion – and treating her as if she were real. Or what about @Azusagakuyuki? Even after 'she' was outed as the very-male Soya, @Azusagakuyuki's follower count continued to climb. At the time of this writing, @Azusagakuyuki has over 33,000 followers – thousands more than when Soya outed himself. The fact that Soya continues to tweet deepfaked pictures of himself as a female shows that his followers don't mind the ongoing facade even though they now know it isn't real.

Yet this ability for an individual to accept the patently obvious and known unreal isn't always without its harms. Just ask Faraz Ansari who obviously knew their lingerie catwalk Reface deepfake was false. Or ask any individual who has been deepfaked into non-consensual porn. Even if the subject and, more pertinently, the viewers of the deepfake know it is false, it can still be humiliating to the person in the deepfake. It's the modern-day equivalent of tarring and feathering someone, though the very real tar and feathers have been replaced with synthetically generated pixels for the crowd's amusement. The deepfake, everyone may know, is false, but the suffering it causes couldn't be more real. Deepfaked pixels can burn just as much as boiling tar.

As I've neared the end of my journey down this rabbit hole of deepfakes, I've increasingly thought back to one of the first deepfakes I ever saw. It's probably one of the first deepfakes many reading this book saw as well. It's a deepfake of Barack Obama created by Oscar-winning filmmaker Jordan Peele in collaboration with BuzzFeed.[134] In it, the deepfaked Obama explains to the audience, 'We're entering an era in which our enemies can make it look like anyone is saying anything at any point in time', before going on to proclaim *Black Panther* antagonist 'Killmonger was right' and that 'President Trump is a total and complete dipshit' to prove the point.

Halfway through the video, it's revealed in a split-screen that Peele himself is voicing the deepfaked Obama, thus pulling back the curtain on the technology and highlighting how duplicitous deepfakes can be. What Peele says on the right-hand side of the frame, the deepfaked Obama continues to speak on the left-hand side. The PSA ends with the 'two' men saying in unison, '[It] may sound basic, but how we move forward, in this age of information, is gonna be the difference between whether we survive or whether we become some kind of fucked up dystopia.'

That warning was from April 2018, and it's even more true today. A year from now, it will be even truer.

Before I began this journey into deepfakes, I had an inkling of a thought, and as I wrote, and researched, and interviewed, that inkling solidified into a decision. A decision to do one final thing.

One final thing that might validate Peele's warning that deepfakes are driving us towards some kind of fucked up dystopia.

One final thing that could reveal if the desire to see something that can't conceivably be real could be so strong that when I finally do see it, its unreality will be irrelevant to me.

I want deepfakes to do the impossible.

I want deepfakes to bring my dad back to life.

Chapter Eight
The End of Death

It was a late, warm September night when I received the page. It was the night of Thursday 3 September, 1998, to be precise. And the reason I received a page and not a phone call was that, back then, few but the well-off had cell phones. I grew up in a middle-class Midwestern suburb of St Louis, Missouri. Both my parents had day jobs and night jobs, and I'd held a part-time job since I was thirteen. I was the only one in my family with a pager, which I paid for from my part-time work at a Subway sandwich shop, and I had one because that's how my group of friends – and many people in their teens and early-twenties where I grew up – communicated in the late 90s.

When I received the page, I was out with my friends. In just two weeks, I'd be moving to Chicago to finish my final two years of college. This would be the college where I'd first be introduced to that Sandra Bullock photo and the world of fake celebrity porn. So, this Thursday night out with my hometown friends would be the last I'd see of them for a while.

I felt my pager buzz and fished it out of my pocket. Like most pagers back in the day, the only information you received on its thin LCD display was the number you were supposed to call back. It was a number I didn't recognise and one that didn't have what we called a 'friend code' after it. Because a

page couldn't contain letters – only numbers and the # and
★ symbols – the only way to know who was paging you was
if you had their phone number memorised. But memorisa-
tion only helped if a friend was paging you to call their home
phone. If that friend was paging you to a payphone or another person's home phone, you wouldn't know who it was
just from the number the pager displayed. That's where 'friend
codes' came in. Everyone who owned a pager chose their own
unique two-digit code which they would append to the end
of the phone number they entered to be called. For example,
my code was ★33, and one of my best friend's code was ★54.
So, even if I didn't recognise the phone number paged to me,
I would know the page was from my best friend by seeing the
'★54' after the number.

But this page I received on that late summer night had no
friend code following the phone number. Besides, it was a little
after nine at night and I was with all my friends. We were at one
of their parent's houses. I asked if I could use the home phone,
and I called the number on my pager's display. It took several
rings before someone answered.

It was my mom. She was paging me from the offices where
she worked her night job, a marketing call centre. She sounded
worried.

'Your dad hasn't come to pick me up yet,' she said.

I grew up in a family of five. It was me, Mom, Dad, my older
brother and younger sister. Though I had the best parents you
could ask for, financially, we didn't have a lot. We were by no
means living in poverty, but money was always tight. It's why
Mom had a day and night job, and Dad did, too. As for my dad's
job, he was a college professor of almost thirty-five years. He
taught business and finance at a local community college – an

honourable profession, yet one that makes it hard to support a family of five on a teacher's salary alone. So, in addition to teaching classes during the day, Dad also taught night courses.

My parents only had one car between them (the other had recently broken down – again), which is why after his daytime classes ended, Dad would drop Mom off at her night job before going to a satellite college forty-five minutes away to teach his night class. When that class ended, he'd drive back to pick up Mom from her work before coming home together.

I looked at my watch. Over the phone, Mom said, 'I got out of work over thirty minutes ago and have been waiting outside for your dad ever since. He's never late.' Now Mom was back inside the offices using a work phone, but security needed to close and lock up the building. 'I'm worried something's wrong.'

'Did you call his office at the school?' I said.

Mom said she had, but no one had picked up.

'Do you want me to come to get you?'

'I know you're out with your friends,' my mom said. I could hear in her voice how bad she felt. She knew it was the last night I'd see them all together. 'Let's give Dad another fifteen minutes. Maybe it's just a flat, and he's on his way. If he doesn't come by then, I'll page you back.'

'OK,' I said. 'I'm sure it's fine. I'm sure he's on his way.'

'I know,' Mom said. We hung up.

Fifteen minutes later, I received another page.

'I'm really sorry,' Mom said when I called her back, 'but they need me to leave the building to lock up, and your dad's still not here. I'm getting really worried now.'

I told her it was fine. I looked at my watch – almost 9.30. Mom's workplace was in an office park, and even though it stays light late on St Louis summer nights, by the time I got there, it'd be dark, and she'd need to be waiting outside since they were locking up. So we made plans for me to pick her up across the street outside a McDonald's. From the safety of the well-lit place, she'd be able to see Dad's car turn into the office park's parking lot and flag him down whenever he finally showed.

On the drive there, I wasn't worried. I figured by the time I arrived, Dad would be there too, late because of car troubles, and I'd drive back to continue the night with my friends. I didn't mind the pointless trip. It was a warm night and I had the top off on my Jeep Wrangler. I loved the feeling of the breeze flowing over me as I travelled down the highway listening to the radio, and the last of dusk's purple skies went black.

I pulled into the McDonald's just after ten. Mom was there, standing outside under the lights. Alone. She looked worried. And now, so was I.

She told me something was terribly wrong because Dad wouldn't let her sit there after work without contacting her. She said he didn't look good when he'd dropped her off. She said he looked tired.

We discussed what to do. Wait to see if he arrived? Or drive the forty-five minutes to where he taught his night courses? We decided on the latter, taking the highways Dad would have taken to pick Mom up. This way, we could keep an eye out for a broken-down car on the other side of the road. My parents were always having problems with it, and if Dad were late picking Mom up, it'd likely be because the car had gone kaput along a stretch of highway where there was no easy way to reach a payphone. That's what I was thinking driving there, anyway. When I looked at Mom, scanning the opposite side of the highway, I felt she was thinking something else.

By the time we arrived, it was almost 11 p.m. and the college buildings were locked up tight. We found campus security and talked to one of the night guards. 'Professor Grothaus left hours ago – right after his class ended,' he told us. I remember the look on Mom's face, hearing those words. I felt a sinking feeling in my stomach.

The guard let us use the security phone to call our house. Maybe Dad hadn't been feeling well, as Mom had said he'd looked, and he accidentally forgot about picking her up and instead went home to go to bed? Yet knowing Dad, that was unlikely. He didn't forget others. Besides, Mom had already tried calling him from her work and even now our home phone just rang and rang until the answering machine picked up.

We discussed what to do. Head home and see if Dad was there but hadn't heard the phone for some reason? Or head back to her workplace? Maybe on the way here, we'd passed him on the highway after he fixed some kind of car trouble and got moving again?

We decided to head back to Mom's workplace, scanning the dark shoulders of the highway in silence, the warm night air whipping over us in the open Jeep. Suddenly, Mom pointed to something just as we passed it. A lone police cruiser parked on the side of the highway. We looked at each other and quickly pulled onto the shoulder. I got out and approached the cruiser, flashing my hands, so the officer inside could see them. He stepped out and followed me to my Jeep, where Mom was now visibly upset.

She explained to the officer how her husband had seemingly gone missing. She explained how the campus security guard said he'd left on time as usual; how we drove this highway and didn't see any stranded car or accident; how we called home and no one answered. Dad had, for all intents and purposes, vanished.

After taking my dad's name and the make and model of his car, the officer walked back to his cruiser. He spoke on the CB unit for several minutes. 'Ma'am,' he addressed Mom when he returned to us, 'someone matching your husband's description was brought to St John's Hospital a few hours ago. They were in an automobile accident.'

If Mom weren't sitting down, she would have collapsed.

The police cruiser pulled off the shoulder. We followed it down the highway as its red and blue bulbs spun and flashed silently in the night. We followed that speeding cruiser all the way to St John's, arriving just after midnight. I'm not sure what happened to the officer. The next thing I remember is walking into the emergency room with my mom crying, 'Where is my husband? Where's my husband?'

What had happened was my dad had finished teaching his night course as normal. He left the campus as normal, too, and headed to pick Mom up. But between the campus and my mom's workplace, Dad passed out at the wheel. It was never certain whether he'd had a heart attack (it would have been his fourth) or if he'd passed out due to some kind of diabetic shock (Dad developed diabetes from a pancreatic infection years earlier). Regardless, when he passed out, he was on the highway and his car drifted into the lane of oncoming traffic and collided with another vehicle head-on.

The impact was so forceful, the engine in my dad's car ended up in its front seats. When emergency crews arrived, firefighters needed to use the jaws of life to saw and cut the mangled steel around Dad's body to extract him from the wreckage. Miraculously, both my dad and the person in the vehicle he collided with survived. Dad was severely cut up; he had a broken wrist and a broken ankle, but he survived, and the occupant of the other vehicle suffered even fewer injuries than him.

Two weeks later, I headed up to Chicago to begin my junior year of college at my new school as planned, thanks to one of my hometown friends offering to drive me the six hours from St Louis to the Windy City (my brother and sister had headed off to their university three hours away weeks before our dad's accident). At my new college, I made friends quickly and enjoyed my classes, and back home, things were going even better. Dad was discharged from the hospital in mid-October. Because of the accident, my parents had lost their last remaining car, but I left them my Jeep, so they had some transportation. Mom felt foolish driving a kitted-out Wrangler with nerf bars, KC floodlights and mudding tires with sixteen-inch chrome rims, but it allowed them to get around as Dad recovered. Things could have been worse.

Then, suddenly, things did get worse.

During the last week of October, Dad felt pain in his leg. Mom took him in my Jeep to the hospital, where he was diagnosed with a blood clot. The clot could have been the result of injuries Dad endured from the accident, though he'd had lifelong cardiovascular and circulation issues. Because of the damage his body had already sustained from the accident, doctors admitted him to the hospital again to treat the clot. Yet within a few days, something else began inside Dad's body. He started retaining fluid at an alarming rate – fifty pounds worth. 'His body looked like it was made of water balloons,' Mom would later tell me. 'Like if you took a needle and pricked him, he would burst.' That fluid retention was a result of the sudden congestive heart failure Dad was experiencing.

241

As the fall progressed, Mom didn't hide his condition from me, but whenever she called I had the feeling she wasn't fully disclosing her worry. When I asked, both she and Dad said things were going well: he was recovering from the congestive heart failure and things were looking up. He might even be out of the hospital by Christmas.

That last part never happened – not fully, anyway. When I came home for Christmas break at the end of 1998, Dad was still in the hospital. However, the doctors gave us permission to bring him home on Christmas Eve and again on Christmas Day so long as he was back before midnight. When I saw him on Christmas Eve, Dad was far from the ballooned fluid-retaining man Mom had described, but he was just as unrecognisable from the neck down. My dad was never really overweight, but he'd always been a thick-framed person. But now? He was physically half his former self. He was skeletal from almost two months of being bedridden. He also couldn't really walk on his own any more; his muscles atrophied. My brother and I had to carry him from the car, across the yard, into the house and onto the couch in front of the Christmas tree. I never knew an adult body could weigh so little. His oxygen canister was heavier than him. Still, we – my mom, brother, sister and I – got to spend Christmas with Dad and, in January, my siblings and I all went back to our respective schools feeling like things were headed in the right direction.

Later that month though, Dad started retaining fluid again – and this time, his condition worsened. He was moved to the intensive care unit where he stayed. All throughout, Mom held the world together with her bare hands, telling her children not to worry, things will be fine – 'Concentrate on your studies. I'm OK. Dad will be OK. Don't worry.' All the while she was driving my Jeep to her day job and then her night job and spending any free time sat at Dad's bedside until hospital rules required her to leave for the night.

And all this while my dad was fighting, too. His body was deteriorating, but Mom said his mind was still as sharp as a tack. He wasn't giving up. And that fight paid off. By the end of February, he was declared well enough to leave the ICU and transferred to a skilled nursing home on the hospital's premises to begin physical rehab. By April, he was back in a regular hospital room again, doing better than he'd been in months.

The last time I would see my dad alive was on Sunday 4 April, 1999. It was Easter Sunday and I was home from college to spend the holiday weekend with my parents at the hospital. Dad was still rail-thin, but one thing that was unchanged was his broad smile and willingness to laugh. He wasn't allowed to leave the hospital, but Mom and I were allowed to take him in a wheelchair down to the 'fancy' Johnathan Livingston Seagull restaurant in the lobby, where the three of us had a hospital food Easter Sunday brunch and talked for hours. I returned that night and sat with Dad in his room until early evening when I had to return to Chicago for classes the following day. I remember kissing him on his cheek as I left, his sharp stubble scratching my face as it always did. Mom would later tell me that after I left, Dad told her it was the perfect day. He was so glad I had come home, even if it was just a short visit.

Sixteen days after that, on 20 April, my dad was deemed well enough to be released from the hospital. It was a day of celebration. For the first time in almost six months, he could go home with Mom and stay there as he continued his recovery. And just nine days later, my dad, it would be decided, was able to do something he'd so hoped he'd be healthy enough to for the longest time. On 29 April 1999, the college where Dad had taught for nearly thirty-five years was having a ceremony for retiring teachers, himself included.

He was so thrilled about this because just a month earlier, the possibility of him being able to go was nothing more than a pipe dream. But his doctor gave him permission, and a friend from my mom's part-time job offered to drive them in her van; my dad was in a wheelchair. Mom said Dad was so excited, not just to be among those honoured at the ceremony for his decades of teaching, but because he was going to see all his colleagues and many friends for the first time in ages in a setting that wasn't a hospital room. Mom helped him get all dressed up in a suit and tie.

'He looked so good,' she told me later. 'He looked like the picture of health.' Then Mom and her friend helped Dad to the van. He kidded around with the two of them as they drove to the college.

They arrived at campus at ten past noon. The ceremony wasn't to start until 1 p.m., but they arrived early, partly, as they needed the extra time to traverse the campus with my wheelchair-bound Dad. But mainly, they arrived early because Dad couldn't wait to get to the student centre, where the ceremony was being held. Many of his colleagues were already inside.

They parked in the west parking lot and got Dad out of the van and into his wheelchair. It was a beautiful day, not a cloud in the sky, as Mom, her friend and Dad made their way across campus towards the student centre.

Then Dad said something.

'The light is so bright.'

Mom's friend looked at her. It was a bright sunny day, but they'd been in the deep shade of a building for several moments.

244

My mom looked at my dad.

'The light is so bright,' he said again.

Then he began to twitch. And twitch.

Mom went cold.

Others on the campus saw what was happening. Someone called security as Mom desperately tried to communicate with her husband. The campus guard who rushed over was friends with him. The campus nurse that arrived shortly after knew the man in the wheelchair as well – she was his distant cousin. Then the paramedics were there. The nurse began unbuttoning my dad's shirt as Mom spoke to him.

'They're going to help you, Glenn,' Mom said. 'I love you, Glenn.'

'I love you, too,' Dad replied.

Those were the last words he would ever speak.

His head slumped back.

The paramedics lifted him from his wheelchair and laid him on the ground. They worked on him for forty minutes, just a short distance from where a student centre full of colleagues were waiting to welcome him back. But that time the paramedics worked on his frail body paid off. Because they didn't quit, they got a pulse and rushed to the closest hospital. Mom stayed at his side throughout, but as the day progressed, and the doctors and nurses came and went, and he remained immobile, she knew Dad was already gone, in a way. His brain had been without oxygen for the forty minutes he'd been unconscious.

It wasn't until after midnight that Mom called me up in Chicago. She'd waited so long because she dared not leave Dad's side during the time the hospital staff tried to stabilise his body and talk through his options – his chances – as he deteriorated beside her. I was in a friend's room in the dorms when one of my roommates came to get me, saying my mom was on the phone. I went down to my room and took the receiver in my hand.

'Michael,' she said. 'You need to get home right away. Your father isn't going to make it.'

It was the middle of the night. A plane ticket was an unaffordable option for my family. And the Chicago–St Louis train wouldn't leave until 10 a.m., and with stops, the trip could take eight hours. As I packed, trying to figure out a way to get home, word had spread around the dorms about what had happened. My friend Abbey came to me. She had a car – a Jeep, too. She offered to drive me the six hours to St Louis. We left at first light.

To this day, I don't remember much about that ride. I'm not sure what Abbey and I talked about. I'm not sure what I thought about. The next thing I do remember is, almost right at noon, pulling up to the hospital. Abbey parked at the curb next to the entrance. She told me she'd wait until I came back out to tell her what I wanted to do. I jumped out of the Jeep and ran inside.

Twenty minutes later, I returned. I thanked Abbey for driving me. I told her it was OK now. She could go. I was staying. My dad died forty minutes before we arrived, just as Abbey and I would have been crossing the Mississippi River into St Louis, at 11.18 a.m. on 30 April 1999.

Abbey made the six-hour return drive to our dorms without me.

In those last moments of my dad's life, Mom was at his side in the hospital, holding his hand. 'Enough is enough,' she told him. 'You have been through enough. It's time for you to go. You have to go. It's OK. It's OK. I will be OK, and the kids will be OK. But you need to go.'

And at those words, Dad passed away. He was only 59.

At the college where he taught, there's a bench dedicated to his memory not far from where he collapsed.

My dad was a moral man. And by that, I don't mean 'religious' (although he was religious, as was I at the time, too). And I definitely don't mean self-righteous, as many religious people I've known are where I grew up. What I mean by 'moral' is my dad was one of those too-rare people who understood that everyone has the right to be treated fairly and compassionately – no matter who they are or what they've done. That's not just my observation of how he lived, that's something I remember him telling me when I was a young teenager.

Though my dad taught as a college professor, he would also sometimes teach business and finance courses at a local prison. I remember finding this out when I was around the age of thirteen and I immediately asked him if he taught robbers. He said yes. I asked him if he taught murderers. He said yes. And next I remember asking him why he was teaching 'bad people who did bad things'. It's a question I still feel so foolish for having asked, but it's one that my dad answered in a way that helped reframe the way I consider people to this very day.

Dad looked at me and said that sometimes people do things out of desperation – things that we might not understand because life isn't the same for all of us. We don't all have the same luck or are granted the same fairness or have the same opportunities. And because of this, he said, some people sometimes have to make hard choices others will never have to. Yes, sometimes those choices may be wrong, but people who have been forced by life to make these hard decisions – and even the people who have done bad things and did them not from unjust desperation, but from pure malice – still deserve to be treated humanely, compassionately and with dignity, too. And none of them should be denied the opportunities that others with more privileged lives and the luxury of easier choices have access to – like access to an education.

'No matter what someone has done, they still deserved a fair shot in life,' Dad said. 'Everyone deserves a second chance – even the so-called "bad" ones.'

From this conversation, I would eventually come to understand one of life's greatest truths: you can only judge the virtue of a society by how it treats the worst among it. The same is true for judging the virtue of an individual. My dad believed the so-called 'worst' among us were as deserving of compassion, opportunity, and second chances as our so-called 'best'. The world today needs more men like my dad. He taught me to look deeper into people and events before making judgements – even when the rest of the world has already issued its condemnation. As a writer and a journalist that lesson has been invaluable.

But that lesson is not the only thing I got from my dad. I get two of my prominent traits from him, too. The first is my love of travel. Before Dad married, he'd been to over thirty countries. And on most trips, he'd also bring the latest consumer technology of the time with him: his Super-8 film camera. I remember, growing up as a kid, on Friday nights, watching the films he shot on his travels before his children were born. My parents and siblings and I sat in the darkened living room eating stove-popped popcorn as the film projector swirled its reels around, projecting the footage onto an unrolled screen. One particular reel I can't get out of my head to this day featured footage Dad filmed of hundreds of monkeys overrunning some small town and monastery somewhere in Southeast Asia. It was a sight so fantastical I had trouble at the time believing it was real, and I would demand to see it replayed – the monkeys running backward as the reels rewound.

The existence of that fantastical footage represents the other prominent trait I get from my dad: my love of technology. Back when Dad was in his twenties, a Super-8 film camera was the most cutting-edge consumer tech you could own. Later, when VHS camcorders became the hottest tech when I was in my youth, Dad would sometimes rent one from a local camera shop (we couldn't afford one outright). And when personal computers came along in the mid-80s, and the public internet came into being in the early 90s, Dad was probably even more excited about those technologies at the time than I was.

My dad died in the final months of the twentieth century, before cell phones, then MP3 players, and then smartphones were commonplace. But whenever I hear about a revolutionary new technology, he's still one of the first people I think about. I suppose that's only normal when you lose someone, and a sufficient amount of time has passed – an amount of time where the world that was when they were alive no longer is, and thus some of the changes that have occurred might seem like magic to them. My dad never could have conceived of any number of the twenty-first-century technologies that have come into being after his passing, but I know he would be captivated by them. Technologies like streaming movies, electric vehicles, GPS navigation, commercial drones, smartwatches with built-in ECGs, real-time video calls via the tiny multitouch computers in our pockets, digital assistants like Siri and Alexa, artificial intelligence, quantum computing that manipulates the fundamental particles of reality itself – and, yes, deepfakes.

And it's the latter I know Dad would find as one of the most fascinating technologies that have come about since his passing. It blends two of his most significant interests: motion pictures and computing.

I know Dad would have loved to play around with deepfake technology if he could have, just like he loved playing around with his Super-8 camera and, later, the first Commodore 64 he bought for our family in the 1980s. (I still remember him sitting in our dining room, where we set the Commodore up, typing away BASIC language commands on the chunky mechanical keyboard.) I also know Dad would have gotten a kick out of the fun face-swap smartphone apps. And I know he would have been amazed at how artificial intelligence can remould the images video has already captured.

And the more I thought about my dad as I talked to all kinds of people involved in the world of deepfakes, I've also come to know one more thing as the writing of this manuscript is coming to a close . . .

'Hey,' Brad says over audio chat. 'I finished your book a few weeks ago. Fuck.'

He's talking about *Epiphany Jones*.

'Is that what Hollywood is really like? I bet that's what it's really like.'

I tell him I'm not wholly sure. Parts of it could be. But I'm a novelist, and that book is fiction and so I made most of the story up. 'It's kind of the job description,' I say.

'Still. *Fuck*. And the last part with his sister and Epiphany in the upstairs room. Fucking made me almost cry. No joke.'

I tell him thanks, although I'm not sure that's the appropriate response.

'So, what do you want this time?'

Brad means, what deepfake do I want now? When I reached him a few days earlier, I asked if he'd be willing to make another deepfake for me. I told him this time I'd pay him because it's personal. But when I tell Brad what I want, he tells me he won't accept payment. He'll do it, but he won't accept payment.

'If it was celeb porn, I'd take your cash,' he says. 'But for this, no.'

I tell him I appreciate that.

'Besides,' he adds, 'I rarely get to do guys' faces. This will just help broaden my deepfaking repertoire.'

We go over the logistics. What I've decided. How to send the required materials to him.

When we're done, Brad says, 'You sure you want this, man?'

But Brad's asking the wrong question. The right question is, 'Are you sure this is a good idea?' That answer is, 'No.'

But my response to Brad's actual question is, 'Yeah, I'm sure.'

'OK,' Brad says, but he says it as if I've told him something else altogether.

It's over a week later when we speak again. Brad tells me my latest deepfake request has been completed.

Truthfully, I'm more nervous to see this deepfake than the one he made of me committing the armed robbery – and the uncertainty of not knowing what type of criminal scenario Brad would deepfake me into was overwhelming. But now I'm so nervous, I don't know what to say.

But Brad goes on. He tells me how the source video I gave him was in standard definition, but the destination video I gave him was in high-definition. He explains how the difference in resolutions could have made the face look blurry, so he scaled down the resolution of the destination video I sent him.

251

'It still looks great. I took extra time on the training. The deepfake just looks like it's old school – VHS quality.'

A decade or so ago, I'd begun digitising some old home videos from my youth so I could save them onto my computer for posterity. But they were digitised at their original VHS resolution. Before I began working on this book, I'd forgotten I had about sixty seconds of digitised footage of Dad. It was a video recorded with one of the VHS camcorders he rented when I was younger. I don't remember the exact date of the original recording, but by the looks of Dad and our kitchen in the video, it was probably shot in the early 90s. That was the video clip I sent to Brad.

'That's fine. VHS-resolution is fine.'

What Brad did with the sixty or so seconds of digitised VHS footage was break it up into almost 1,800 individual images of Dad's face. He then ran those images through DeepFaceLab on his 'rig', as he calls it, to train a model of my dad's face. Brad then deepfaked my dad, at my request, onto the body of a person in a videoclip I sent him, too.

'Well, that's the link.' Brad says. I look at the encrypted message he just sent. He tells me he stripped the metadata from the video. 'It's a direct download. Once you download it, I'll erase it from my server. And just so you know, I've already deleted his faceset and model.'

'I appreciate that,' I tell him.

'Don't worry about it.' Then, 'So until you need another deepfake . . . I guess this is goodbye.'

I let out an involuntary breath of laughter. 'I'm done with deepfakes after this.'

'Well then, good luck with the book. Hopefully, it's as good as the novel.'

'Hopefully,' I say.

Before I hang up, I ask, 'Hey. Will you let me send you a copy of the hardcover when it's out? As a thank you?'

Brad is quiet for a moment. 'Nice try, man. But, nah. I like my anonymity, and I'm not giving out my address.'

It's not until later that night that I finally sit down at my laptop and open the video file. For some reason, it feels easier watching it knowing most of the world around me has been tucked into bed for hours already. It's as if all that exists on this night is me and the video – the deepfake – I'm about to watch of Dad.

My hand feels tingly as I hover my finger over the trackpad. Then I press it. On my laptop's screen, the video's play head fades away. Instantly, my dad appears. It's him. One hundred per cent, it's him. There are his bushy eyebrows and his thick moustache and full cheeks, and his broad forehead with its receding greying hairline.

Dad's wearing a light-yellow polo shirt and is on a trail in a leafy green park. The shot is framed from his chest on up, with one arm outstretched towards the frame. The reason for this is because Dad is holding a device in his hand, one that didn't exist during his lifetime. He's holding a modern-day smartphone, and he's recording a video of himself with it.

And as he's recording this selfie, he's slowly inching backwards, smiling into the smartphone's camera, using the device's large display in his hand to make sure the shot is framed so I can see the beautiful park behind him. After a second, Dad shifts his head to glance out of frame and he gives a friendly smile. A moment later, a jogger with a dog in tow comes into frame on the opposite side of the path Dad's moving down. The jogger turns her head and gives him a pleasant nod, before receding down the path into the distance.

Dad turns back towards the camera and smiles and then brings his free hand – the one not holding the smartphone – up into frame. He gives me an affectionate wave as he continues to inch backwards down the path. He then stretches that hand back like a tour guide, leading my line of sight to the beautiful splotches of light created from the gaps in the trees overhead whose branches stretch across the path like a canopy. The splotches of light sway gently over the benches under the trees.

Then Dad turns back towards me one last time and, for the final moments of the video, he simply looks into the camera and smiles a nice warm smile at me. He waves one more time with that big smile on his face and, as he looks at me, the video fades to black.

I watched this deepfake of my dad a dozen more times – immediately after I finished the first viewing. And then a dozen more. And right now, I could write another dozen pages, explaining the technical specifics of the deepfake. I could explain how I scoured the web days before contacting Brad again. I could explain how I searched to find a video that featured a man of Dad's build and with a greying receding hairline that would match Dad's perfectly, but wasn't having any luck. I could explain how I didn't want to leave this video up to chance – or have any more surprises by letting Brad pick it.

I could explain all of this.

But none of that matters right now.

What matters most is I was able to view a deepfake allowing me to see my dad doing something I know he would have loved. Something he missed out on due to his passing too soon: using a pocket-sized twenty-first-century computer to record a movie – just as he used the most advanced tech of his day to record his travel films.

And that was important to me not solely so that I could write about it for this book. The fact is, I just wanted to see my dad again – for me. For myself. Because it's now been twenty-two years since he died. Twenty-two years since he was taken too soon. And twenty-two years later, the technology now exists that allows me to see him – and see him again in a way that existing authentic video of my dad can't provide. Authentic video of Dad – it's a *historical* record. It's the him in the video taken in our kitchen in the early 90s. It's something that already happened and that, yes, I can repeatedly watch to my heart's content, but it will never be a record of something new, something unexpected – something that makes it feel as if Dad is out there, today, alive and being recorded afresh.

And so, that night, on my couch, as the world all around me slept, I watched the deepfake of my dad one more time. I saw him do something new one more time.

And then I deleted the deepfake.

We human beings are a deeply flawed species. Many of those flaws stem from the fact that we are not always rational. So often, we make choices based on emotion, not logic. Emotion led me to decide that it was OK to use deepfake technology to let me see my dad again. But what that emotion compelled me to do next is what made me realise I needed to delete the deepfake then and there.

Because that emotion was telling me one thing: 'You need to show this to Mom.'

And that would have been wrong.

That last time I talked to my mom about Dad's death, she said something about the experience that stuck with me. She said that whole time – from the night of his accident on 3

September 1998, to the day of his death on 30 April 1999 – 'It was an odyssey.'

An odyssey.

There were the horrible parts, and so many of them. The dread she felt after we found the police officer on the side of the highway and hearing him say a man matching her husband's description had been in an accident. The touch and go nature of Dad's congestive heart failure that followed, watching his body go from ballooned to rail-thin and back again. All the while, Mom trying to temper any worries her children may have had – desiring for them to not agonise about their father's condition. All this while Mom went daily from her morning job to her evening job, heading to the hospital every free moment she had in-between, sitting at her husband's bedside for as long as she was allowed. Then, her husband's collapse at the most unexpected time and place – all after the worst was supposedly behind them. And finally, needing to tell the love of her life that it was OK – he had suffered enough – and it was OK for him to go now. Those were the words he needed to hear before he would allow himself to leave this world – and her.

But during those months, Mom also says she experienced profound moments of awe. These moments were what she called 'god winks', and many of them came in the form of the kindness of strangers. Random people gave her money to help pay the bills or they brought her food because she so often didn't have the time to stop to eat between work and hospital. Sometimes a complete stranger in the hospital just happened to say the right words to her when she needed to hear them most – people who had no idea about her husband's circumstances. These were just some of the fortuitous things that occured during that time to help her get through – the serendipitous coincidences that shouldn't have happened but did.

Then, most importantly, there was, of course, all the time she got to spend with her husband at his hospital bedside. The time they had together to talk about their lives, the ups and downs, and reminisce about everything they had been through – all their shared experiences, the good and the bad. That time they had during those eight horrible months brought them closer than ever before.

An odyssey, indeed.

And I didn't want to take that away from her. I didn't want to cheapen her last memories of my dad and all they had been through by showing her a synthetic recreation of him doing something new – 'living' in the modern world again.

That's why I deleted the deepfake from my laptop then and there.

I've thought about the deepfake of Dad every day between the night I deleted it and now as I'm writing these words. Of course, Dad wouldn't have cared I deepfaked him. As a matter of fact, I'm certain he would have loved it and been amazed. Yet that fact doesn't help answer an important question – and it's a question only the living can answer. With deepfake technology, we now have it in our power to see our loved ones anew, long after they've passed. But should we use that power?

That is something we all must answer for ourselves. I could write pages and pages about the potential use of this technology for 'good' – how it could be used to bring closure to people; how sufferers of dementia could 'FaceTime' their long-dead spouse, bringing back some semblance of the world they've forgotten doesn't exist any more. I could go on about all the theoretical possibilities in the medical and healthcare industries. But I won't.

The technology is there – and I hope you've learned enough from reading this to know just what can be achieved. But we do have to pause from time to time and ask ourselves: just because we can, does that mean we should? I can't answer that for anyone else, and I'm not convinced there is a right or a wrong answer.

But as for me? Here's what I believe.

What happened, happened. You can't change the past. And after having tried it myself, after having digitally raised the dead, I can speak from experience: the dead should stay dead. Reviving them ties the living to a synthetic shell of what our loved ones once were. In the process, it stops the living from moving forward by trapping them to the past.

And that's no way to spend the time we have left.

But of course, you might feel differently.

Virtually every journalist, scientist and researcher hopes their dire predictions will one day be lambasted as 'dramatic', 'proven wrong' or 'unfulfilled'. That's part of the reason those same people write books – books like this. The intent is to warn what can happen if we do nothing. The hope, of course, by voicing their concerns, be it in a book or a research paper, is that the dire consequences they foresee can be avoided – that their predictions don't turn out to be brilliant foresight but alarmist overreaction. Yet, it will only be overreaction if their warnings spur individuals, institutions and governments to prepare for the threat before it overruns our world.

And that is my hope for this book. There is no doubt deepfake technology has profound positive use cases in fields ranging from entertainment to education to healthcare. And as a tool of creative expression and an entirely new medium of art, deepfakes' potentials are nearly limitless. But as we've seen from our exploration throughout this book, deepfake technology also has the potential to radically alter our world for the worse. It is already being used to harm and humiliate a significant number of people. In the coming years, its capacity for suffering will only advance from being capable of injury at the individual level to causing damage on a global scale. And, ultimately, if we are left unprepared, deepfake technology has the potential to erode the most cherished and fundamental binding material human societies rely on – interpersonal trust.

Deepfakes are already influencing our lives today – right this moment – and they can even seemingly alter the past, as I've shown. But no deepfake will ever be powerful enough to fix our future. Deepfakes will undoubtedly be able to show us a utopian tomorrow even if our world becomes 'some kind of fucked up dystopia', as Jordan Peele's deepfaked Obama warned, but such deepfakes will simply be a type of wonderful synthetic digital opiate that we can anaesthetise ourselves with from time to time to get through our bleak days.

It is only we, in the present, who can act to protect our future, and the time to begin is now.

Notes

1   Bash, J. & Steed, M. (2020). 'Deepfakes threaten the 2020 election'. *The Hill.* https://thehill.com/opinion/cybersecurity/508202-deepfakes-threaten-the-2020-election

2   Finney Boylan, J. (2018). 'Will Deep-Fake Technology Destroy Democracy?' *New York Times.* https://www.nytimes.com/2018/10/17/opinion/deep-fake-technology-democracy.html

3   Wolfgang, B. (2018). 'Vladimir Putin's 'deep fakes' threaten U.S. elections'. *Washington Times.* https://www.washingtontimes.com/news/2018/dec/2/vladimir-putins-deep-fakes-threaten-us-elections/

4   Kent, T. (2018). 'Opinion: Fake news is about to get so much more dangerous'. *Washington Post.* https://www.washingtonpost.com/opinions/fake-news-is-about-to-get-so-much-more-dangerous/2018/09/06/3d7e4194-a1a6-11e8-83d2-70203b8d7b44_story.html

5   Waddel, K. (2019). 'The 2020 campaigns aren't ready for deepfakes'. *AXIOS.* https://www.axios.com/2020-campaigns-arent-ready-for-deepfakes-a1506e77-6914-4e24-b2d1-0c69b6e22162.html

6   Lima, C. (2019). "Nightmarish": Lawmakers brace for swarm of 2020 deepfakes'. *Politico.* https://www.politico.com/story/2019/06/13/facebook-deep-fakes-2020-1527268

7   Harwell, D. (2019). 'Top AI researchers race to detect "deepfake" videos: "We are outgunned"'. *Washington Post.* https://

www.washingtonpost.com/technology/2019/06/12/top-ai-researchers-race-detect-deepfake-videos-we-are-outgunned/

8   Watts, C. & Hwang, T. (2020). 'Deepfakes are coming for American democracy. Here's how we can prepare'. *Washington Post*. https://www.washingtonpost.com/opinions/2020/09/10/deepfakes-are-coming-american-democracy-heres-how-we-can-prepare/

9   George, S. (2019). "Deepfakes" called new election threat, with no easy fix'. *AP.* https://apnews.com/article/nancy-pelosi-elections-artificial-intelligence-politics-technology-4b8ec588bf5047a981bb6f7ac4acb5a7

10   Shao, G. (2019). 'Fake videos could be the next big problem in the 2020 elections'. *CNBC.* https://www.cnbc.com/2019/10/15/deepfakes-could-be-problem-for-the-2020-election.html

11   Lasky, S. (2018). 'U.S. Intel agencies warn about Deepfake video scourge'. *Security InfoWatch.* https://www.security-infowatch.com/video-surveillance/video-analytics/article/12422323/us-intel-agencies-warn-about-deepfake-video-scourge

12   Roby, K. (2019). 'Education and legislation are needed to combat the significant threat of deepfakes'. *TechRepublic.* https://www.techrepublic.com/article/the-sinister-timing-of-deepfakes-and-the-2020-election/

13   Burroughs, T. (2019). 'Deepfake Videos Set to Wreak Havoc'. *Issues.* https://issues.org/news/deepfake-videos-set-to-wreak-havoc/

14   Goodfellow, I. J., Pouget-Abadie, J., Mirza, M., Xu, B., Warde-Farley, D., Ozair, S., Courville, A. & Bengio, Y. (2014). 'Generative Adversarial Networks'. Cornell University. https://papers.nips.cc/paper/2014/file/5ca3e9b122f61f8f-06494c97b1afccf3-Paper.pdf

15   Face Swap Live website. See: http://faceswaplive.com

16   Dewey, C. (2016). 'Behind the scenes of Face Swap Live, the "creepy" app that launched a thousand memes'. *Washington Post*. https://www.washingtonpost.com/news/the-intersect/wp/2016/01/13/behind-the-scenes-of-face-swap-live-the-creepy-app-that-launched-a-thousand-memes/?variant=116ae929826d1fd3

17   Apple Newsroom. (2016). 'Apple unveils Best of 2016 across apps, music, movies and more'. Apple.com. https://www.apple.com/newsroom/2016/12/apple-unveils-best-of-2016-across-apps-music-movies-and-more/

18   Miller, M. (2017). 'This Actress Secretly Played Princess Leia in *Rogue One*'. *Esquire*. https://www.esquire.com/entertainment/movies/news/a53856/rogue-one-princess-leia-actress/

19   Derpfakes. (2018, Jan 27). *Princess Leia CGI | Deepfakes Replacement*. [Video]. YouTube. https://www.youtube.com/watch?v=614we6ZaQ04

20   Lomas, N. (2017). 'FaceApp apologizes for building a racist AI'. *TechCrunch*. https://techcrunch.com/2017/04/25/face-app-apologises-for-building-a-racist-ai/

21   Harwell, D. (2019). 'Federal study confirms racial bias of many facial-recognition systems, casts doubt on their expanding use'. *Washington Post*. https://www.washingtonpost.com/technology/2019/12/19/federal-study-confirms-racial-bias-many-facial-recognition-systems-casts-doubt-their-expanding-use/

22   Reface App. See: https://hey.reface.ai

23   Raj, R. (2020). 'Andreessen Horowitz backs Ukrainian face-swap app Reface's €4.53M seed round; find out more about this round'. *Silicon Canals*. https://siliconcanals.com/news/startups/reface-raises-4-53m/

24   Chiu, K. (2020). 'Censorship sweep punishes deepfake app Reface and online school Xueersi in China's latest internet crackdown'. *South China Morning Post*. https://www.scmp.com/

tech/article/3106284/censorship-sweep-punishes-deepfake-app-reface-and-online-school-xueersi-chinas

25   'The top 500 sites on the web'. Alexa.com. https://www.alexa.com/topsites

26   Skitz4twenty YouTube Page. See: https://www.youtube.com/c/Skitz4twenty/

27   Durrati. (2018). 'Wow! Marjory Stoneman Douglas High Survivor Emma Gonzalez Takes It to Trump, GOP and NRA. UpDated.' *Daily KOS.* https://www.dailykos.com/stories/2018/2/17/1742324/-Wow-Marjory-Stoneman-Douglas-High-Survivor-Sophomore-Emma-Gonzales-Takes-It-to-Trump-GOP-and-NRA

28   Gonzalez, E. (2018). 'Emma Gonzalez on Why This Generation Needs Gun Control'. *Teen Vogue.* https://www.teenvogue.com/story/emma-gonzalez-parkland-gun-control-cover

29   Danner, C. (2018). 'People are Sharing Fake Photos of Emma González Tearing Up the Constitution'. *NY Magazine Influencer.* https://nymag.com/intelligencer/2018/03/some-conservatives-are-sharing-a-fake-photo-of-emma-gonzalez.html

30   Harwell, D. (2019). 'Faked Pelosi videos, slowed to make her appear drunk, spread across social media'. *Washington Post.* https://www.washingtonpost.com/technology/2019/05/23/faked-pelosi-videos-slowed-make-her-appear-drunk-spread-across-social-media/

31   Relman, E. (2019). 'Rudy Giuliani tweeted a doctored video of Nancy Pelosi minutes before Trump attacked Pelosi with another misleading video'. Business Insider[itals]. https://www.businessinsider.com/rudy-giuliani-tweets-doctored-video-nancy-pelosi-2019-5

32   Dupuy, B. (2020). 'Video of a Joe Biden speech was misleadingly edited to suggest he made a racist remark – and then went viral'. *CNBC.* https://www.cnbc.com/2020/01/02/

joe-biden-video-edited-to-suggest-racist-remark-went-viral.html

33 Javers, E. (2020). 'Facebook and Twitter decline Pelosi request to delete Trump video'. *CNBC*. https://www.cnbc.com/2020/02/07/facebook-and-twitter-decline-pelosi-request-to-delete-trump-video.html

34 Zakrzewski, C. (2020). 'Twitter flags video retweeted by President Trump as "manipulated media"'. *Washington Post*. https://www.washingtonpost.com/technology/2020/03/08/twitter-flags-video-retweeted-by-president-trump-manipulated-media/

35 Kharpal, A. (2020). 'Trump's "racist baby" tweet gets slapped with "manipulated media" label from Twitter'. *CNBC*. https://www.cnbc.com/2020/06/19/trump-racist-baby-tweet-gets-slapped-with-manipulated-media-label.html

36 Samuels, E. (2020). 'White House social media director tweets manipulated video of Biden'. *Washington Post*. https://www.washingtonpost.com/politics/2020/09/02/white-house-social-media-director-tweets-manipulated-video-biden/

37 Haberman, M. [@maggieNYT]. (2020, Aug 31). *Harry Belafonte on the manipulated video featuring him tweeted by Scavino, a dep WH chief of staff…* [Tweet]. Twitter. https://twitter.com/maggieNYT/status/1300516131996872704

38 Wiegel, D. (2020). 'Twitter flags GOP video after activist's computerized voice was manipulated'. *Washington Post*. https://www.washingtonpost.com/politics/2020/08/30/ady-barkan-scalise-twitter-video/

39 Barkan, A. [@AdyBarkan]. (2020, Aug 30). *@SteveScalise, These are not my words. I have lost my ability to speak, but not my agency or my thoughts…* [Tweet]. Twitter. https://twitter.com/AdyBarkan/status/1300159116942274560

40 Hains, T. (2020). 'Trump Campaign Highlights Obviously Out-Of-Context Biden Quote: "You Won't Be Safe In Joe Biden's

America'". *RealClear Politics*. https://www.realclearpolitics.com/video/2020/08/31/trump_campaign_highlights_obviously_out_of_context_biden_quote_you_wont_be_safe_in_joe_bidens_america.html

41 Swenson, A. (2020). 'Video altered to make it look like Biden greeted wrong state'. *AP*. https://apnews.com/article/joe-biden-video-altered-58124115393828f85cd496514bba4726

42 Johnson, D. [@TheRock]. (2020, Sep 27). *As a political independent & centrist, I've voted for both parties in the past. In this critical presidential election…* [Tweet]. Twitter. https://twitter.com/therock/status/1310198847835000834?lang=en

43 Nicolaou, E. (2020). 'What Made Judy Garland's Life So Tragic'. *Refinery29*. https://www.refinery29.com/en-us/2019/09/8455229/judy-garland-actress-true-story-life-death

44 Cole, S. (2017). 'AI-Assisted Fake Porn is Here and We're All Fucked'. *VICE*. https://www.vice.com/en/article/gydydm/gal-gadot-fake-ai-porn

45 SimilarWeb. 'Top Website Ranking'. See: https://www.similarweb.com/top-websites/

46 'The 2019 Year in Review'. *PornHub*. https://www.pornhub.com/insights/2019-year-in-review

47 Buchholz, K. (2019). 'How Much of the Internet Consists of Porn?'. *Statista*. https://www.statista.com/chart/16959/share-of-the-internet-that-is-porn/

48 The Editors. (2017). 'What We Learned About Sexual Desire From 10 Years of Pornhub User Data'. *The Cut*. https://www.thecut.com/2017/06/pornhub-data-sexual-habits.html

49 'The Pornhub Tech Review'. (2021). *PornHub*. https://www.pornhub.com/insights/tech-review

50 Cole, S. (2018). 'We Are Truly Fucked: Everyone Is Making AI-Generated Fake Porn Now'. *VICE*. https://www.vice.com/en/article/bjye8a/reddit-fake-porn-app-daisy-ridley

51 DeepFaceLab. GitHub. https://github.com/iperov/DeepFaceLab

52 Sinola, V. (2018, Feb 5). *Cage will survive*. [Video]. YouTube. https://www.youtube.com/watch?v=dh-QM54RuAs

53 (2020). 'How long is the typical film actor's career?'. *Stephen Follows*. https://stephenfollows.com/how-long-is-the-typical-film-actors-career/

54 Dewey, C. (2016). 'Is Rule 34 actually true?: An investigation into the Internet's most risqué law'. *Washington Post*. https://www.washingtonpost.com/news/the-intersect/wp/2016/04/06/is-rule-34-actually-true-an-investigation-into-the-internets-most-risque-law/

55 Cole, S. (2019). 'This Horrifying App Undresses a Photo of Any Woman With a Single Click'. *VICE*. https://www.vice.com/en/article/kzm59x/deepnude-app-creates-fake-nudes-of-any-woman

56 [@deepnudeapp]. (2019, Jun 27). *Here is the brief history, and the end of DeepNude. We created this project for user's entertainment a few months ago...* [Tweet]. Twitter. https://twitter.com/deepnudeapp/status/1144307316231200768

57 Cavalli, F. (2020). 'Automating Image Abuse: deepfake bots on Telegram'. *Sensity*. https://sensity.ai/automating-image-abuse-deepfake-bots-on-telegram/

58 Ayyub, R. (2018). 'I Was The Victim Of A Deepfake Porn Plot Intended To Silence Me'. *HuffPost*. https://www.huffingtonpost.co.uk/entry/deepfake-porn_uk_5bf2c126e4b0f32bd58ba316

59 Ayyub, R. (2018). 'In India, Journalists Face Slut-Shaming and Rape Threats'. *New York Times*. https://www.nytimes.com/2018/05/22/opinion/india-journalists-slut-shaming-rape.html

60 Umawing, J. (2020, updated 2021). 'The face of tomorrow's cybercrime: Deepfake ransomware explained'. *Malwarebytes Labs*. https://blog.malwarebytes.com/ransomware/2020/06/the-face-of-tomorrows-cybercrime-deepfake-ransomware-explained/

61   Harwell, D. (2019). 'An artificial-intelligence first: Voice-mimicking software reportedly used for major theft'. *Washington Post*. https://www.washingtonpost.com/technology/2019/09/04/an-artificial-intelligence-first-voice-mimicking-software-reportedly-used-major-theft/

62   Protalinski, E. (2012). 'Chinese spies used fake Facebook profile to friend NATO officials'. *ZD Net*. https://www.zdnet.com/article/chinese-spies-used-fake-facebook-profile-to-friend-nato-officials/

63   This Person Does Not Exist. https://thispersondoesnotexist.com

64   https://www.patreon.com/lucidrains

65   Karras, T., Laine, S. & Aila, T. (2019). 'A Style-Based Generator Architecture for Generative Adversarial Networks'. 4396-4405. 10.1109/CVPR.2019.00453. https://arxiv.org/pdf/1812.04948.pdf

66   @DFRLab. (2021). 'Inauthentic Instagram accounts with synthetic faces target Navalny protests'. *DFRLab*. https://medium.com/dfrlab/inauthentic-instagram-accounts-with-synthetic-faces-target-navalny-protests-a6a516395e25

67   Mirsky, Y., Mahler, T., Shelef, I. & Elovici, Y. (2019). 'CT-GAN: Malicious Tampering of 3D Medical Imagery using Deep Learning'. Usenix Security Symposium. https://www.usenix.org/system/files/sec19-mirsky_0.pdf

68   Virginia's Legislative Information System: 2019 Session. House Bill No. 2678. See: https://lis.virginia.gov/cgi-bin/legp604.exe?191+ful+HB2678S1+hil

69   Texas Gov: An Act relating to the creation of a criminal offense for fabricating…'. See: https://capitol.texas.gov/tlodocs/86R/billtext/html/SB00751F.htm

70   California Legislative Information: Assembly Bill No. 730. See: https://leginfo.legislature.ca.gov/faces/billTextClient.xhtml?bill_id=201920200AB730

71 California Legislative Information: Assembly Bill No. 602. See: https://leginfo.legislature.ca.gov/faces/billTextClient.xhtml?-bill_id=201920200AB602

72 Maryland General Assembly. Statutes Text. See: https://mgaleg. maryland.gov/mgawebsite/Laws/StatuteText?article=gcr&-section=11-208&enactments=false

73 The New York State Senate. Senate Bill S5959D: 2019-2020 Legislative Session. See: https://www.nysenate.gov/legisla-tion/bills/2019/s5959

74 Harwell, D. (2018). 'Scarlett Johansson on fake AI-generated sex videos: "Nothing can stop someone from cutting and pasting my image"'. *Washington Post*. https://www.washingtonpost.com/technology/2018/12/31/scarlett-johansson-fake-ai-generated-sex-videos-nothing-can-stop-someone-cutting-pasting-my-image/

75 Greene, D. (2018). 'We Don't Need New Laws for Faked Videos, We Already Have Them'. *EFF*. https://www.eff.org/deeplinks/2018/02/we-dont-need-new-laws-faked-videos-we-already-have-them

76 ViralHog. (2014, Sept 23). *Attempted Robbery Caught on GoPro.* [Video]. YouTube. https://www.youtube.com/watch?v=a-78JsiEXS2Y

77 Mithaiwala, M. (2019). 'Disney Owns 8 of the 10 Biggest Movies Of All Time'. *Screen Rant*. https://screenrant.com/dis-ney-owns-highest-grossing-movies-worldwide/

78 Vary, A. B. (2020). 'Disney Explodes Box Office Records With $11.1 Billion Worldwide for 2019'. *Variety*. https://variety.com/2020/film/box-office/disney-global-box-of-fice-2019-1203453364/

79 Smith, C. (2021). 'Disney Revenue Statistics and Details | By the Numbers'. *DisneyNews*. https://disneynews.us/disney-rev-enue-statistics/

80   Goldsmith, J. (2021). 'Disney+ Tops 100 Million Subscribers – CEO Bob Chapek'. *Deadline*. https://deadline.com/2021/03/disney-plus-tops-100-million-subscribers-1234710077/

81   Traverse, C. (2020). 'Who is Virtual Influencer, Model, and Digital Fashion Designer Aliona Pole?'. *Virtual Humans*. https://www.virtualhumans.org/article/who-is-virtual-influencer-model-and-digital-fashion-designer-aliona-pole

82   Takumi. (2020). 'TAKUMI Launches Into the Mainstream: Influencer Marketing in Society Whitepaper'. *LBB Online*. https://www.lbbonline.com/news/takumi-launches-into-the-mainstream-influencer-marketing-in-society-whitepaper

83   Rahal, A. (2020). 'Is Influencer Marketing Worth It In 2020?'. *Forbes*. https://www.forbes.com/sites/theyec/2020/01/10/is-influencer-marketing-worth-it-in-2020/?sh=8096f9731c54

84   Imma. [@imma.gram]. Instagram Profile. See: https://www.instagram.com/imma.gram

85   Malivar website. See: https://malivar.io/en

86   AI Face Replacement: pinscreen. See: https://www.pinscreen.com/facereplacement/

87   Synthesia website. See: https://www.synthesia.io

88   Synthesia Case Study – David Beckham / Malaria No More / RGA. (2020). See: https://www.synthesia.io/post/case-study-david-beckham-malaria-no-more-rga

89   Revoice website. See: https://www.projectrevoice.org

90   New York State Assembly. Bill No. A08155. See: https://nyassembly.gov/leg/?default_fld=&leg_video=&bn=A08155&term=2017&Summary=Y&Actions=Y&Floor Votes=Y&Memo=Y&Text=Y

91   Cole, S. (2017). 'AI-Assisted Fake Porn Is Here and We're All Fucked'. *VICE*. https://www.vice.com/en/article/gydydm/gal-gadot-fake-ai-porn

92   Naruniec, J., Helminger, L., Schroers, C. & Weber, R. M. (2020). 'High-Resolution Neural Face Swapping for Visual Ef-

fects'. Disney Research Studios, Eurographics Symposium on Rendering. https://studios.disneyresearch.com/2020/06/29/high-resolution-neural-face-swapping-for-visual-effects/

93   Open letter from The Walt Disney Company to Senator Martin Golden opposing A.8155-b. (2018). See: https://www.rightofpublicityroadmap.com/sites/default/files/pdfs/disney_opposition_letters_a8155b.pdf

94   Bass, J. (2018). 'New York Right of Publicity Bill Passage Drama Ends With No action by State Senate'. The Entertainment, Arts and Sports Law Blog. http://nysbar.com/blogs/EASL/2018/06/new_york_right_of_publicity_bi.html

95   Memorandum in Opposition to New York Assembly Bill A.8155B. (2018). Motion Picture Association of America, Inc. See: https://www.rightofpublicityroadmap.com/sites/default/files/pdfs/mpaa_opposition_to_a8155b.pdf

96   Memorandum in Opposition to New York Assembly Bill A08155B (Right of Publicity). (2018). NBC Universal. See: https://www.rightofpublicityroadmap.com/sites/default/files/pdfs/nbc_opposition_a8155b.pdf

97   Memorandum in Opposition to Assembly Bill 8155-B, Senate Bill 5857-B (Right of Publicity). (2018). Electronic Frontier Foundation. See: https://www.eff.org/files/2018/06/08/eff_memorundum_in_opposition_to_a8155b_and_s5857b.pdf

98   The New York State Senate. Senate Bill S5857B. 2017-2018 Legislative Session. See: https://www.nysenate.gov/legislation/bills/2017/s5857/amendment/b

99   New York State Assembly. S05959 Summary. See: https://www.nyassembly.gov/leg/?default_fld=&leg_video=&bn=S05959&term=2019&Summary=Y&Actions=Y&Committee%26nbspVotes=Y&Floor%26nbspVotes=Y&Memo=Y&Text=Y&LFIN=Y

100   'In Event of Moon Disaster'. MIT. https://moondisaster.org

101 Safire, W. & Nixon, R. (1969). 'Statement for President Nixon to read in case the astronauts were stranded on the Moon, July 18, 1969'. National Archives and Records Administration, U.S. Government Printing Office. See: https://en.wikisource.org/wiki/In_Event_of_Moon_Disaster

102 Canny AI website. See: https://www.cannyai.com

103 Respeecher website. See: https://www.respeecher.com

104 Keller, B. (1989). 'Major Soviet Paper Says 20 Million Died As Victims of Stalin'. *New York Times.* https://www.nytimes.com/1989/02/04/world/major-soviet-paper-says-20-million-died-as-victims-of-stalin.html

105 King, D. (1997). *The Commissar Vanishes: The Falsification of Photographs and Art in Stalin's Russia.* (New York: Metropolitan Books).

106 Christopher, N. (2020). 'We've Just Seen the First Use of Deepfakes in an Indian Election Campaign'. *VICE.* https://www.vice.com/en/article/jgedjb/the-first-use-of-deepfakes-in-indian-election-by-bjp

107 newsfromindiahere. (2020, Feb 18). *MT Original.* [Video]. YouTube. https://www.youtube.com/watch?v=2Tar2O4q0qY

108 newsfromindiahere. (2020, Feb 18). *MT English.* [Video]. YouTube. https://www.youtube.com/watch?v=88GUbuL89bQ

109 newsfromindiahere. (2020, Feb 18). *MT Haryanvi.* [Video]. YouTube. https://www.youtube.com/watch?v=ZAdrE_wEMM0

110 Mihindukulasuriya, R. (2020). 'Why the Manoj Tiwari deepfakes should have India deeply worried'. *The Print.* https://theprint.in/tech/why-the-manoj-tiwari-deepfakes-should-have-india-deeply-worried/372389/

111 Breland, A. (2019). 'The Bizarre and Terrifying Case of the "Deepfake" Video that Helped Bring an African Nation to the Brink'. *Mother Jones.* https://www.motherjones.com/politics/2019/03/deepfake-gabon-ali-bongo/

112 Africanews with AFP. (2019). '"I am now fine": Ali Bongo tells Gabonese in New Year message'. *africanews*. https://www.africanews.com/2019/01/01/i-am-now-fine-ali-bongo-tells-gabonese-in-new-year-message/

113 Gabon 24. Discours A La Nation Du Président Ali Bongo Ondimba. *Facebook*, 31 Dec 2018. Accessed 28 Jul 2021. See: https://www.facebook.com/watch/?v=324528215059254

114 Hallam, M. & Pieper, O. (2020). 'Germany: Online child abuse investigators to get more powers'. *DW*. https://www.dw.com/en/germany-online-child-abuse-investigators-to-get-more-powers/a-52037583

115 Turek, M. Media Forensics (MediFor) page. DARPA. See: https://www.darpa.mil/program/media-forensics

116 Congress.gov. S.1790 – National Defense Authorization Act for Fiscal year 2020. 116th Congress (2019-2020). See: https://www.congress.gov/bill/116th-congress/senate-bill/1790/actions

117 宗谷の蒼氷. [@azusagakuyuki]. (2020, Jul 18). *It's the best!!* (translated from Japanese by Google). [Tweet]. Twitter. https://twitter.com/azusagakuyuki/status/1284437921584443393

118 宗谷の蒼氷. [@azusagakuyuki]. (2020, Aug 9). *3XV9 Revival Daisakusen 23…* (translated from Japanese by Google). [Tweet]. Twitter. https://twitter.com/azusagakuyuki/status/1292408391298478083

119 宗谷の蒼氷. [@azusagakuyuki]. (2021, Feb 11). *Ah 3XV9!? Really…* (translated from Japanese by Google). [Tweet]. Twitter. https://twitter.com/azusagakuyuki/status/1359812725220675600

120 Hernandez, P. (2018). 'For Women on Twitch, Disclosing Their Relationship Status is a Minefield'. *The Verge*. https://www.theverge.com/2018/8/8/17661596/twitch-relationship-status-amouranth-women-donations-single

121 (2021). 'Japan's "beauty iron knight" loved by thousands, reveals that he is an uncle'. *Singtao*. https://www.singtao.

ca/4827354/2021-03-16/news-日本「美女鐵騎士」萬千寵
愛+++真身曝光原來是個大叔/?variant=zh-hk

122 (2021). 'Face editing: Japanese biker tricks internet into thinking he is a young woman'. *BBC*. https://www.bbc.co.uk/
news/world-asia-56447357

123 Ewe, K. (2021). 'Young Female Japanese Biker Turns Out To
Be 50-Year-Old Man With FaceApp'. *VICE*. https://www.
vice.com/en/article/y3g88m/viral-japanese-biker-transfor-
mation-woman-faceapp

124 Ovadya, A. & Bienstock, H. (2018). 'Is Your Company Ready
to Protect Its Reputation from Deep Fakes?' *Harvard Business
Review*. https://hbr.org/2018/11/is-your-company-ready-to-
protect-its-reputation-from-deep-fakes

125 Chesney, R. & Citron, D. K. (2019). 'Deep Fakes: A Looming
Challenge for Privacy, Democracy, and National Security). 107
California Law Review 1753, U of Texas Law, Public Law Re-
search Paper No. 692, U of Maryland Legal Studies Research
Paper No. 2018-21, Available at SSRN: https://ssrn.com/ab-
stract=3213954 or http://dx.doi.org/10.2139/ssrn.3213954

126 Derysh, I. (2020). 'GOP House candidate publishes lengthy report claiming George Floyd's killing was a "deepfake" hoax'.
Salon. https://www.salon.com/2020/06/24/gop-house-can-
didate-publishes-lengthy-report-claiming-george-floyds-kill-
ing-was-a-deepfake-hoax/

127 Sonnemaker, T. (2021). '"Liar's dividend": The more we
learn about deepfakes, the more dangerous they become'.
Insider. https://www.businessinsider.com/deepfakes-liars-
dividend-explained-future-misinformation-social-me-
dia-fake-news-2021-4

128 Reuters Staff. (2019). 'China seeks to root out fake news and
deepfakes with new online content rules'. *Reuters*. https://
www.reuters.com/article/us-china-technology-idINKB-
N1Y30VU

129 Bankhead III, M. (2021). 'How to spot deepfakes? Look at light reflection in the eyes'. *University at Buffalo News Center*. http://www.buffalo.edu/news/releases/2021/03/010.html

130 'What Is Blockchain-Based Timestamping and Who Needs It?'. Origin Stamp. https://originstamp.com/blog/what-is-blockchain-based-timestamping/

131 Ibrahim, M. (2020). 'To beat deepfakes, we need to prove what is real. Here's how'. World Economic Forum. https://www.weforum.org/agenda/2020/03/how-to-make-better-decisions-in-the-deepfake-era/

132 Caulfield, M. (2019). 'SIFT (The Four Moves)'. HAPGOOD. https://hapgood.us/2019/06/19/sift-the-four-moves/

133 FBI Cyber Division. (2021). 'Malicious Actors Almost Certainly Will Leverage Synthetic Content for Cyber and Foreign Influence Operations'. Private Industry Notification, Federal Bureau of Investigation, Cyber Division. https://www.ic3.gov/Media/News/2021/210310-2.pdf

134 Mack, D. (2018). 'This PSA About Fake News From Barack Obama Is Not What It Appears'. *BuzzFeedNews*. https://www.buzzfeednews.com/article/davidmack/obama-fake-news-jordan-peele-psa-video-buzzfeed

Acknowledgements

This book would not have been possible without the kindness and support of a significant number of people whose input and generosity have helped inform and shape the final work.

To my editors, Harriet Poland and Izzy Everington, who guided me on my journey, helping me see things I was blind to.

To Phillipa Geering, my producer on this book's associated podcast, who conducted hours of interviews with me that were also critical in the creation of this text. Her perspectives broadened mine.

To Dominic Gribben and Philip Abrams, who executive produced the podcast and provided crucial editorial insight and feedback.

To Giorgio Patrini, CEO of Sensity. Both he and his company are among the best repositories of knowledge about deepfakes on the planet.

To David Greene, Senior Staff Attorney and Civil Liberties Director of The Electronic Frontier Foundation, one of the twenty-first century's most important civil liberties organizations and a critical defender of journalistic freedoms, free expression, and digital privacy. David's work protects some of our most essential liberties, and I'm grateful for the time and insight he lent me.

To Rebecca A. Delfino, Clinical Professor of Law at Loyola Marymount University, whose research into deepfakes predates mine and who is a fount of wisdom regarding their nuanced legal implications and challenges.

To Faraz Ansari, whose films are as brave as themself.

To Kristina Sweet aka 'Luxury Girl' for her fascinating and insightful perspectives on the impact deepfakes are having on the business professionals and performers in the porn industry.

To the talented people at Malivar and Aww Inc. for their preview of the future in the form of Aliona and Imma.

To Christopher Travers, a brilliant business leader on a mission to develop the virtual human space. His breadth of knowledge on the subject is astounding.

To Mike Caulfield, the creator of the SIFT framework. I'm thankful for his feedback and the fact that he has the foresight to think of ways to mitigate emerging informational threats before they overwhelm us.

To Skitz4Twenty. *Dude.*

To Luke Dormehl for his tidbits of wisdom.

To the late historian and archivist David King whose incredible book *The Commissar Vanishes, The Falsification of Photographs and Art in Stalin's Russia* helped contextualize for me the historical horrors of media manipulation. Media and history buffs must check out his work.

To Halsey Burgund, a gifted artist whose work on altering the past has shown us a glimpse of our future.

To the team at Respeecher, whose mind-blowing audio voice cloning tech shows precisely why deepfake technology has such transformative potential for the entertainment and communication industries.

To Ryan Laney, VFX extraordinaire and a great human being whose conversation was a gift. Though not mentioned in this book, Laney's work using deepfake technology to enable

LGBTQ+ peoples in Chechnya to tell their stories to the world while shielding their identities in the excellent documentary *Welcome to Chechnya* is one of the best examples of deepfaking for good you'll ever come across.

To Samantha Cole and all the other journalists, experts, and researchers whose exploration of and work on deepfakes helped make this book possible.

To my mom for her strength in recounting my dad's story – and for being an amazing mom.

To anyone else I have failed to mention as I rush to make my publisher's deadline.

Finally, to the people in these pages who asked to keep their real identities private yet told me their stories with honesty in order to help me better tell the story of deepfakes.